30. 55
Scientia
11 Oct 1995

NATURAL PHILOSOPHY AT DARTMOUTH

NATURAL PHILOSOPHY
AT DARTMOUTH
From Surveyors' Chains to
the Pressure of Light

BY

Sanborn C. Brown AND *Leonard M. Rieser*

Distributed for Dartmouth College

by

THE UNIVERSITY PRESS OF NEW ENGLAND

Hanover, New Hampshire 03755

1974

CONTENTS

Preface   ix

Chapter 1. A Professor of Natural Philosophy
             Helps Found a College   1

         2. A Lonely Scholar   15

         3. Natural Philosophy Drifts   26

         4. Administrative Turmoil   34

         5. Science Can Be Exciting   44

         6. For the Glory of God   57

         7. The Coming of Research   68

         8. "A General Utility Man"   85

         9. A One Man Department No Longer   100

        10. The End of an Era   112

            Epilogue   121

            Index   123

ILLUSTRATIONS

1. Bezaleel Woodward.   7

2. Solution to a Problem.   8

3. Surveyors' Chain.   9

4. Semi-Circumferentor.   10

5. The English Telescope.   13

6. Ticknor's Fat Little Professor.   24

7. A Problem in Trigonometry.   30

8. John Hubbard.   32

9. The Orrery.   36

10. Ebenezer Adams.   41

11. Ira Young.   48

12. Ira Young's Observatory.   50

13. Early Picture of the Observatory.   55

14. Henry Fairbanks.   61

15. Astronomers at Burlington, Iowa.   73

16. The Tele-Spectroscope.   73

17. Charles A. Young.   81

18. Illustration for a Textbook.   94

19. Another Illustration.   95

20. Charles Emerson.   98

21. The Wilder Physical Laboratory.   99

22. Vacuum Tubes.   99

23. The Dartmouth Crookes Tube.   108

24. The Nichols and Hull Experiment.   108

25. Ernest Nichols.   120

PREFACE

AT the present time colleges and universities in the United States are facing a redirection which may change their whole educational pattern. Particularly in science, after a period of unprecedented support for both pure and applied research as a necessary adjunct to the teaching function, finances are being withdrawn and the search for knowledge for its own sake is being questioned. Science is being blamed for many of society's ills and the teaching of science is being discouraged and downgraded. Should the universities be the fountainhead of pure science research, or should the professors go back full time to their classes and let the industrial and government laboratories, with their demonstratively greater "cost effectiveness," be the ones that are supported to extend the boundaries of knowledge?

The increasing difficulty being experienced by professors of science and technology to raise funds for their research is forcing the educational community in this country into a reassessment of the role of the college teacher and the place of research and scholarship in the academic scene. This is by no means the first such revolution that has occurred in American higher education and looking at models of the past can be useful in deciding directions for the future.

Physics is in the focus of this attack. Government support for university research in this area is being cut. Research projects are being shifted from private universities to government-owned laboratories. Applied physics money is being funneled into industrial organizations and the universities are being told to go back to their teaching and leave research to others.

This book is a reminder of what happens to the educational process when scholarship is discouraged and the search for new knowledge is unsupported. It is a series of historical sketches chosen in the area of physics, or "natural philosophy" as it used to be called, of interest now because this is the subject under the sharpest question at present. It is not a history of physics but a history of the development of a research commitment as presented through the lives of a series of 18th and 19th century physicists held together by the thread of professional continuity as teachers in the same college. It traces the evolution of what we call today a "science department" in a small New England college from its missionary-to-the-Indian founding when its only experimental work was learning to survey the land, to its establishment of a research tradition in our modern sense with the experimental proof of Maxwell's equations and the verification of the existence of X rays.

One characteristic of this study will be amply clear to the reader. This account is not presented against the background of what was going on at other comparable colleges in the United States. It is an account almost entirely of physics at Dartmouth alone. This emphasizes the point that those were the days when there was in fact almost no communication between institutions. Professors did not get together to discuss their mutual problems; colleges hired their own graduates to maintain their isolation and particularly in those institutions far from cities; and time spent in traveling was considered ill-used compared with the full-time commitment to teaching and working for the college. Except where noted toward the end of the era covered by this volume, Dartmouth College was little affected by what was going on in neighboring institutions in New England.

This book started out to be a narrative history of the physics department at Dartmouth College, but after a few weeks of work it became obvious to us that a much more interesting issue was involved than the simple growth of a science department in a small New England college. Not only the content of physics, but the whole approach to physics as an intellectual discipline and its interaction with society has changed profoundly in the last 200 years. Now as the forces of society seem to want to push science into the background again, and skeptical looks are cast on new knowledge and pure research, our small sample of professors take on the role of models from history as to what science education might once more become. If one wants to de-emphasize research and scholarship in the university setting, one can go backward through this series of professors and choose which stage one finds ideal. It is in this sense that we see our study as models from history.

We are profoundly grateful to a host of persons who have given generously of their time, effort, and particular skills. The staff of Dartmouth's Baker Library not only turned itself upside down to produce all the material we needed for this study, but the comfortable and unique surroundings of wood-paneled studies far from telephones and other disturbances and yet equipped with all the necessary dictating machines and secretaries sped this task on its appointed way in a manner which is hard to over-appreciate.

Mr. Kenneth C. Cramer, the Dartmouth College Archivist, must surely have worked full time for us while we were in our two writing sessions, since the material flowed from his department in a continuous stream. More than that, during the years when we were not together in Hanover his prompt replies to our many enquiries kept the project moving ahead.

Professor Julius A. Brown (father of one of the authors), a physicist and astronomer who had his training with all of the later figures in this book, joined us for a few days in Hanover to guide us, with his firsthand knowledge, through the wealth of material about the late nineteenth century physicists and astronomers. We found that when we took him with us to interview people about their recollections, having an octogenarian along as part of the team enhanced the flow of information tremendously.

We believe that the illustrations in the book are outstanding, and this results from the skill, patience, and understanding of the Dartmouth College photographer, Adrian N. Bouchard. We are most grateful for all his help.

Mr. Edward C. Lathem, Librarian of Dartmouth College, and Professor Allen L. King of the Physics Department have read drafts of the manuscript and their suggestions and criticisms have been most appreciated.

Anyone who has written a book knows of the debt one always owes to one's secretaries, who never seem to have the fun, but so much of the hard work. In this case we have been unusually fortunate in this part of our labors, and it is a pleasure to express our thanks to Mrs. Stacia Ballou, Miss Barbara Kellerup, Mrs. Jo Fosso, and Miss Doris McEntee.

Sanborn C. Brown
Leonard M. Rieser

Chapter 1

A PROFESSOR OF NATURAL PHILOSOPHY
HELPS FOUND A COLLEGE

For those of us familiar with college life today, the rigors of an 18th-century New England college would remind us more of a penal institution than a place for fitting the elite of the land to become intellectual leaders of society. Dartmouth in the 1770's and 1780's was typical of other American colleges in the rigidity of its curriculum and the austerity of its life. Every student followed an identical course of study, regardless of his tastes, interests, or future profession. For three years two thirds of his time was devoted to Latin and Greek and one third to arithmetic, English grammar, logic, geography, geometry, trigonometry, algebra, conic sections, surveying, mensuration, natural and moral philosophy, and astronomy. In the senior year the classics were discontinued, and the student turned to metaphysics, theology, and "politic law," with much attention to composition and public speaking.

In the early years of Dartmouth College the professors, who had come up through much the same kind of curriculum as they taught, were often capable of teaching many different subjects; but they were usually carried on the rolls of the College as teachers of specific disciplines. One finds Bezaleel Woodward as the first professor of mathematics and natural philosophy not only teaching those subjects but also performing a host of other duties, some of which had fallen naturally on his shoulders as President Eleazar Wheelock's right-hand man even before the College was founded.

Eleazar Wheelock started his educational leadership by founding Moor's Charity School in Lebanon, Connecticut, where young Bezaleel studied before he went on to Yale, from which institution he was graduated in 1764. In the early years Wheelock had customarily sent his graduates first to Princeton and later to Yale. In 1767 six or seven of his pupils, including his son John, were at Yale. Both to curtail expenses and to keep the whole course of education under his own jurisdiction, Wheelock organized a collegiate branch of his Moor's Charity School in 1768. Bezaleel Woodward, a graduate of Wheelock's school who had been out in the countryside as a preacher, returned in November 1766 as a bookkeeper and assistant, and in 1767–68 he acted as a Preceptor of the school. He became a tutor for the new collegiate branch. Thus, when Wheelock's plans for enlarging his educational horizons resulted in the founding of Dartmouth College in 1769 and its establishment in Hanover, New Hampshire, in 1770, it is hardly surprising that Bezaleel Woodward followed him into the wilderness of the North Country and remained throughout his whole life a bulwark of strength and devotion to the struggling college. One notes in passing that this devotion was not only to the College as an institution, since three years after Dartmouth was founded the first marriage recorded in Hanover was between Bezaleel Woodward and Mary Wheelock, a daughter of the president.

To understand in any real way the educational process in colonial America, one must look with some detail into the accepted routine of student and teacher alike. In the early days of Woodward's professorship, the College assembled in the chapel at five o'clock in the morning, or in the winter "as early as the president could see to read the Bible." No artificial lights were provided, and perhaps more significantly no heat was provided either. Hanover, New Hampshire, wakes up not infrequently to a morning of temperatures many degrees below zero, and one can readily imagine the sleepy undergraduates of that era on such mornings wrapped in their warmest clothing, huddled together in an endeavor to keep as warm as circumstances would permit.

At the close of the religious exercises, the students were

summoned to their first recitation—the freshmen in one room, the sophomores in another, and the juniors and seniors each in their respective rooms. The College provided the rooms, but the furnishings were supplied by the students from their own pockets. A chair and table were required for the instructor together with a small slate blackboard, a stove and a double row of unpainted pine benches. What heat and light might be required were also provided by student funds or labor.

After these first two events of the day the students were allowed to go to breakfast, which was followed by a period of study and recitations at eleven and twelve o'clock. Another period of study occurred after lunch. Afternoon classes were held at 3:00 and 4:00. Evening prayers came at six o'clock "or as late as the president was able to see." Time for recreation during the summer was limited to the periods between the close of the morning exercise and eight o'clock, between the close of the late morning recitation and two o'clock, and between evening prayers and nine o'clock. In the winter the schedule was somewhat modified because of the lack of daylight hours, but the total free time was about the same.

At all other periods not specified in this schedule the student was supposed to be in his room and at work. Members of the instructional force were required to visit each room to make sure that its occupant was properly busy and that the work was being conscientiously done. No classes were held on Sunday, but all undergraduates attended two sessions of church, in addition to the usual morning and evening chapel services. Students were strictly forbidden to leave their rooms on the Lord's Day except for religious exercises and for food; "nor could they do business, nor walk abroad, nor make indecent noise or clamor." To make sure that there was no temptation for sinful study on the Sabbath, no morning recitation was held on Monday. Instead, a Biblical exercise was held after breakfast, the preparation for which was not considered to be a misuse of Sunday. One must realize that this rigid schedule had to be adhered to by both faculty and students. In addition to this, the outside duties of a man of Bezaleel Woodward's stature were staggering to consider.

Woodward early acquired legal responsibilities in the community. In January of 1771 the Royal Governor of New Hampshire, John Wentworth, appointed Eleazar Wheelock a justice of the peace, but Wheelock's other duties proved too exacting to allow him to pay the proper attention to this important station. On his request, Tutor Woodward received a similar appointment in June 1772. He was also nominated in 1773 to be one of the justices of His Majesty's Inferior Court of the County of Grafton, and until his death in 1804 he presided in one capacity or another over important local tribunals in both civil and criminal cases. His records were kept most methodically and are even today a model of efficiency and care.

During the American Revolution even further duties were loaded upon this competent man. One finds that at the annual town meeting of March 1775 a vote was passed which reads, "that we highly approve of the measures entered into by the American Continental Congress" and that in conformity with this approval there should be set up "a committee of inspection to see that the said association agreement be faithfully and carefully observed in this town and that the same persons be a committee of correspondence in behalf of the town." The membership included Bezaleel Woodward, and he served as clerk until the committee was dissolved. He exercised more power and influence than any other single person connected with it, and to him the Provincial congress gave fit recognition by reappointing him in January 1776 to the same position of justice that he held by Crown appointment prior to the outbreak of hostilities.

Mr. Woodward took charge of the College whenever the president was away and also during the period following Eleazar Wheelock's death until Eleazar's son John could be persuaded to accept the presidency. He was treasurer of the college for twenty-three years, librarian from 1773 to 1777, and a trustee from 1773 until his death. At a time when the College was in desperate need of funds, he acted as manager of one of the College lotteries. He was a member of the committee which erected the village church. In 1784 the trustees voted to solicit subscriptions for the construction of the building which subsequently became known as Dartmouth Hall, and Bezaleel Woodward was appointed contractor and business manager of the undertaking.

With characteristic enterprise, he set about erecting a sawmill, digging a cellar on the site selected, and obtaining stone for the foundation. Although lack of money prevented actual construction for two more years, it is characteristic of Woodward's incredible energy that he could add this duty to the exacting routine of the required College schedule, along with his other responsibilities.

He was said to be the antithesis of President Eleazar Wheelock. "There was nothing scholastic about his appearance or manners; he was of plain and informal manners and mingled in society as other men who had no connection with the College. He was more popular with the students as a man" than anyone else on the faculty. It is curious that, in all the estimates of Bezaleel Woodward, no mention is ever made of his teaching ability. However, poor teachers have always been conspicuous targets of student scorn, and it is natural to suppose that a man so popular with students and whom they described as clear-headed and unpretentious must have been an excellent teacher, even though it is sometimes difficult to imagine how he found time for his pedagogical duties. His official title was that of Professor of Mathematics and Natural Philosophy. There is no record of what he taught in the field of natural philosophy, but a few records of his course called "a system of plane trigonometry" were preserved by students who carefully wrote up their notes as a memento of their college days.[1]

These copied notes serve as an interesting illustration of the rigidity of the teaching methods of the day. It appears that the Hon. Bezaleel Woodward dictated precisely the same format year after year. Our two examples come from 1801 and 1804, and every word and every example is identical. The values of each sine, cosine, and tangent are carried out to eight significant figures without variation. This was typical of the curriculum

1. The Dartmouth College Archives have manuscript copies of "A System of Plain Trigonometry by the Hon. Bezaleel Woodward, Professor of Math. & Nat. Phil. Dartmouth University," transcribed from the course as given in 1801 and 1804. One was written out by Asa Peabody, who was not graduated, and the other by Samuel A. Kimball of the Class of 1806, whose notes are most complete.

presented to college students throughout the Colonies. No variation was acceptable, and the students were required to recite in the literal sense of the word, and without deviation, the wisdom which their tutors and professors imparted. This practice may well account for the ability of a man like Woodward to do so many things in addition to the teaching of his classes, since, once prepared, he himself probably copied from his own notebook onto the blackboard the material which was to be carefully and faithfully reproduced, in turn, annually in his students' notebooks. Our modern requirement that a good teacher be sufficiently versed in peripheral material to answer any reasonably germane question was completely foreign to the rote-memory technique of the 18th century.

The only variation that appears in Professor Woodward's system of plane trigonometry is in the part called "miscellany," or what we would term "problems": "A gentleman has a farm of one thousand (1000) acres lying in a circle. His house is in the middle of this farm. He divides his farm between his wife and three daughters, giving to his daughters three as large circles as can be obtained in his farm and to his wife the remainder. Each of the daughters is supposed to have a house in the middle of her portion. How much land does each daughter possess and how much the old lady? What distance are the daughters from their father's house and how far from each other?" This was obviously an independent problem to be worked out by the student on his own, and the number of acres contained in the farm does vary from year to year. So also does the formality of the arrangement of the solution. The layout of the problem is illustrated in Figure 2.[2]

Knowing the various duties and responsibilities of President Wheelock and his hard-working lieutenant, we can guess that Bezaleel Woodward shouldered the responsibility of surveying Dartmouth College land. The surveyor's chain and semicircumferentor illustrated in Figures 3 and 4 have been proudly

---

2. The particular example of the gentleman with the circular farms is taken from the notebook of Samuel Ayer Kimball preserved in the Dartmouth College Archives. The drawing appeared on page 27 of this notebook.

1. Bezaleel Woodward, 1745–1804. Original at the home of William and Mary Magenau, 132 Willard Street, New Haven, Connecticut. Said to have been painted by Joseph Steward.

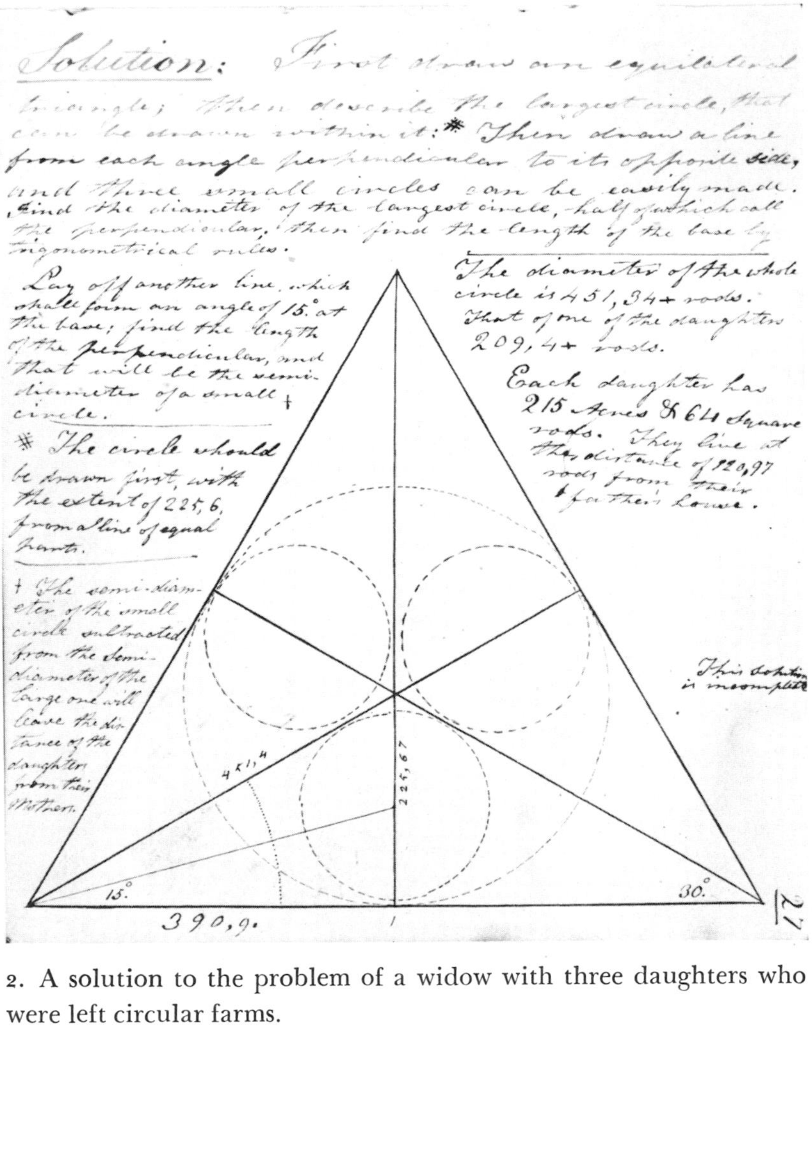

Solution: First draw an equilateral triangle; then describe the largest circle, that can be drawn within it:* Then draw a line from each angle perpendicular to its opposite side, and three small circles can be easily made. Find the diameter of the largest circle, half of which call the perpendicular; then find the length of the base by trigonometrical rules.

Lay off another line, which shall form an angle of 15° at the base; find the length of the perpendicular, and that will be the semi-diameter of a small† circle.

* The circle should be drawn first, with the extent of 225,6, from a line of equal parts.

† The semi-diameter of the small circle subtracted from the semi-diameter of the large one will leave the distance of the daughter from their mother.

The diameter of the whole circle is 451,34+ rods. That of one of the daughters 209,4+ rods.

Each daughter has 215 Acres & 64 square rods. They live at the distance of 120,97 rods from their father's house.

This solution is incomplete

451,4

226,67

15°

30°

390,9.

1

22

2. A solution to the problem of a widow with three daughters who were left circular farms.

3. Surveyor's chain and semi-circumferentor. The chain is a half-chain of 33 feet and the semi-circumferentor is one of three known in this country.

4. The semi-circumferentor.

exhibited in the Dartmouth College Museum as having belonged originally to Eleazar Wheelock, but all the evidence of the division of activities between these two men points to Bezaleel Woodward as the actual surveyor. He was very much interested in the process, and the vast majority of the trigonometry problems that Dartmouth students slaved over in the late 18th century involved the areas and positions of land and the exact locations of houses thereon. One cannot but wonder just which college project may actually have been involved in the exercise when "a person undertakes to remove earth, his employer gives 75 cents for every cube of six feet which he removes. He digs a place into a side hill one hundred and twelve (112) feet horizontally and seventy five (75) feet in width. The ascent of the hill is forty five (45) degrees. How many cubic masses of six feet dimension has he removed and how much money is due him?"

How available published books on trigonometry were to the students of Dartmouth at this time we really cannot discover, but we are sure of at least one. "The unknown sides and angles of a plane triangle are investigated by applying a circle to some one of the sides by which application new terms arise which Mr. Pike calls the 'tabular sides,' others 'the word unit,' its relation to a circle, and so forth" refers to a widely used system of arithmetic published in 1788 by Nicolas Pike. First among the "Recommendations" at the front of this book we find:

> Dartmouth University A.D. 1786
> At the request of Nicolas Pike, Esq. we have inspected his System of Arithmetic, which we cheerfully recommend to the public as easy, accurate and complete. And we apprehend there is no treatise of the kind extant, from which so great utility may arise to Schools
> B. Woodward, Math, and Phil. Prof.
> John Smith, Professor of the Learned Languages

> I do most sincerely concur in the preceding recommendation.
> J. Wheelock, President of the University

We might point out in passing that besides receiving such an encomium from Dartmouth, Pike obtained similar statements from the presidents of Harvard and Yale, the governor of

Massachusetts, and the president of the American Academy of
Arts and Sciences.[3]

While Bezaleel Woodward was still a tutor, Eleazar Wheelock tried to strengthen the area of Natural Philosophy by seeking funds to purchase scientific equipment. In 1772 he turned to
a frequent patron of the college, a Mr. John Phillips, who was
shortly to become a trustee of the College and is well-known today as the founder of the famous academy in Exeter which bears
his name. Mr. Phillips was persuaded to donate 175 pounds to
assist in procuring philosophical apparatus. This donation passed
through many vicissitudes, arising primarily from the fact that
there were no instrument makers of note in America.[4] In May
of 1773 Wheelock put the money in the hands of the Royal Governor, John Wentworth, to arrange for the equipment to be
built in England (Figure 5). In September the Governor gave
the money to his cousin, Paul Wentworth of London, and in the
autumn of 1774 Wentworth informed the Governor that the
instruments were being built "under the inspection of Dr. Sol-
ander and Dr. Irvine by Ramsey who is incomparably the best
hand in Europe." "I have," he said, "mentioned the subscription to Lord Dartmouth who has promised me his support.
These things shall be shipped in the fall ship or in the early
spring with something of my own toward a library." In writing
to Wheelock, Governor Wentworth adds, "I think we may rely
on having the most perfect apparatus and a liberal addition to
Colonel Phillips' generous donation. I shall not cease to study
the interest of Dartmouth College as the best service to the prov-
ince." The shipment was apparently not ready for the fall ship
and could not be sent in the following spring, since open hostili-

---

3. Nicholas Pike's book was entitled *A New and Complete System of
Arithmetic, composed for the citizens of the United States.* The first edition
was published in Newburyport by John Mycall in 1788 and went through
many more editions. From the similarity between the diagrams and se-
quence of problems and theorems found in Pike's book and those copied
from Woodward's class presentation, it appears that the good professor
relied heavily on this book as a source.

4. The most complete documentation of the complicated Phillips donation for "philosophical apparatus" is to be found in Frederick Chase's
History of Dartmouth College (Cambridge, Mass., 1891), pp. 253, 573, and
587.

5. The English telescope which the Wheelock brothers arranged to be given to Dartmouth after more than ten years of negotiation.

ties broke out in April 1775, and the Governor himself fled from the rebellious Colonies.

As matters turned out, the College received this gift twice. After the Revolution was over, John Wheelock went to England to re-establish contacts with the English trustees and other friends. He met with Paul Wentworth, who was perfectly willing to honor his previous commitment, and lined up another donor, a Dr. William Rose of Chiswick, to help pay for the collection. Although nothing really seems to have resulted from this contact by John Wheelock, his two brothers, James and Eleazar, Jr., went to England in 1784 to try to obtain the apparatus. It actually arrived in 1785.[5]

Impatient with all these delays, the College authorities had in the meantime decided that Wentworth was not going to make good his promise, and they therefore petitioned the New Hampshire General Court to reimburse them, with interest, from the proceeds of the confiscated estate of Governor Wentworth. This the legislature did in its own good time, settling the affair finally in 1789. By that time the philosophical apparatus from England had already been in use for three years, and the money seems to have been applied toward the building of Dartmouth Hall.

Bezaleel Woodward was no scholar, but in the early days of the college intellectual attainment was by no means necessary. The rigidity of the curriculum and the rote learning by the students required a stern disciplinarian and a total commitment, night and day, to the college and the "boys." Woodward was eminently successful in both these areas and he typified excellence among late 18th-century New England college professors.

---

5. The list of equipment here attributed to the Wheelocks' trip to England assumes that all the equipment owned by the College in 1787 came from this source. It was in October of that year that the Rev. Ezra Stiles recorded the details of "Dartm. Apparatus.," on p. 396 of *Extracts from the Itineraries and Other Miscellanies of Ezra Stiles, DD., LL.D. 1755–1794*, edited by F. B. Dexter and published by Yale University Press in 1916. This assumption may be only partially justified. Chase in his history (note 4, above) mentions on p. 574 that "two large globes had been received . . . from Mr. Wentworth." He did not give the source of his information. These same globes were lent to Mr. John Hubbard in 1789 when he was preceptor of the New Ipswich Academy.

Chapter 2

A LONELY SCHOLAR

ONE of the striking features of the "Recommendations" printed in the front of Nicolas Pike's *New and Complete System of Arithmetic* was that a "John Smith, Professor of Learned Languages" should be considered a good reference for a system of arithmetic. If this seems strange to us today, it shows our lack of understanding of the educational hierarchy in New England colleges.

The teaching staff consisted of three categories: tutors, professors, and the president. Tutors were supposed to teach a particular class while a professor taught a particular subject. Thus at Dartmouth when there were two tutors available, one was assigned to the freshmen, teaching essentially all the subjects studied by that class, while the other performed a similar duty for the sophomores. The dignified and honorable responsibility of guiding the senior class was entirely in the hands of the president. This left the professors the task of instructing juniors and giving public lectures in their particular fields of learning. The majority of professors at this early period held other offices with the college and might even be tutors at the same time they were professors. The president and the tutors carried out most of the classroom work of the college. The tutors not only had to do with the two lower classes, the most difficult to control, but were also required to room "in college," to board in the Commons, to exercise constant watchfulness over college discipline, and to be constantly subject to the annoyances of exuberant youth.

John Smith, who had been graduated from Dartmouth in 1773, was slender and remarkably pale as a young man. He was very nearsighted and totally unable to take a joke. The students made life thoroughly miserable for him; yet he was the only early Dartmouth professor with any pretensions to productive scholarship. During his lifetime he published a Latin, Greek, and Hebrew grammar, wrote a Chaldaic grammar which was never published, and published an edition of Cicero's *De oratore* with notes and a brief memoir of Cicero in English. It was said that he clearly understood Sumerian, as well as Chaldaic, and that his researches in Arabic were of great extent.

Professor Smith was easily flustered and an irresistible mark for ridicule by the undergraduates. The sport which the students made of him was legendary, and many stories about him are to be found in histories of Dartmouth College. One morning as he was crossing the green in a dense fog he mistook the charred tree stumps on the common for live animals and rushed into the College shrieking that he was being chased by a bear and three cubs. For the rest of his life he was never allowed to forget his moment of terror.

Although the students took advantage of John Smith, his industry and uncomplaining patience seem beyond human powers. After the death of theology Professor Ripley, no professor of this subject was appointed, and the students were thus no longer receiving lectures in the area. The deficiency was considered highly lamentable, and Smith was asked to fill the void. He was the pastor of the churches on both sides of the river (in Hartford, Vermont, and Hanover, New Hampshire). He was the college librarian. He was carrying an extra class in Hebrew and he had the duty of correcting all the exercises that were spoken on the stage. Despite all these demands, he agreed to undertake a set of lectures of which the manuscripts still survive, and for two years he delivered one of these each Sunday evening at college prayers.

In a manuscript description of his life,[1] his wife wrote, "It is almost impossible to have a correct view of the disadvantages

---

1. Deposited in the Dartmouth College Archives.

under which these labors were performed without having witnessed them; his only study for many years was a small room, which was constantly occupied by his family and all the company they had to entertain, which was by no means few in number, but amidst all these hinderances he sat at his desk with his attention immovably fixed upon the studies which he happened to be employed in, as if alone in the world, and here he wrote and rewrote everything he published. . . .

"As an instance of Mr. Smith's indefatigable industry and perseverance when he thought it was for the improvement of others, after he had supposed he had completed his Greek grammar, and taken it to Boston with the expectation of putting it immediately in the press; but on submitting it to the inspection of his friend and fellow townsman Judge Parsons, who was then supposed to be the best scholar in Greek language in the United States, he proposed to him to remodel it upon a new, and as he thought greatly improved plan much superior to any then in use, and promised him if he would comply with his request, he would use all his influence to get it introduced into Cambridge University and it was said in that day, that he had more influence with the literati connected with that institution than any man in the state, and from his suggestions of the advantage it would be to the student in that particular study, Mr. Smith returned home, and patiently sat down to the task of rewriting and remodeling the whole work and finished it about three months before he died; Judge Parsons dying soon after, the influence was never exerted and the work has shared the fate of most of the publications of that day, in being superseded by living authors on their own plans."

The terms under which John Smith agreed in 1777 to become Dartmouth's first professor spell out conditions of personal servitude beyond the wildest nightmares of academic life today:

> An agreement between the Reverend Doctor Eleazar Wheelock, president of Dartmouth College, and Mr. John Smith, late tutor of the same, with respect to said Mr. Smith's settlement and salary in capacity of Professor of the languages in Dartmouth College.
>
> Mr. Smith agrees to settle as Professor of English, Latin,

Greek, Hebrew, Chaldee, etc., in Dartmouth College, to teach which, and as many of these and other such languages as he shall understand, as the Trustees shall judge necessary and practical for one man, and also to read lectures on them, as often as the president, tutors, etc., with himself shall judge profitable for the Seminary. He also agrees, while he can do it consistently with his office as professor, annually to serve as tutor to a class of students in the College. In consideration of which, Dr. Wheelock agrees to give to him (the said Mr. Smith) one hundred pounds L. My. annually as a salary to be paid one half in money and the other half in money or in such necessary articles for a family as wheat, Indian corn, rye beef, pork, mutton, butter, cheese, hay, pasturing etc., as long as he shall continue professor as aforesaid . . . Doctor Wheelock also agrees that Mr. Smith's salary, viz: one hundred pounds annually, shall not be diminished when his business as professor shall be so great, that it will render it impracticable for him to serve as a tutor to a class in College . . . Mr. Smith also promises that whenever he shall have a sufficient support from any fund established for the maintenance of a professor of language, he will give up the salary to which this agreement entitles him."[2]

It may seem strange in a chronicle history of natural philosophy to consider at such length the life of a professor of languages who nowhere in the official lists of Dartmouth teachers appears connected with such technical subjects as physics and mathematics. Yet our only information on the details of the courses and textbooks[3] used in natural philosophy under the first

2. The full agreement between John Smith and Eleazar Wheelock is given in B. P. Smith, *The History of Dartmouth College* (Boston, 1878), pp. 452–453.

3. Examining textbooks is one of the most illuminating ways of discovering course content. The books in Natural Philosophy in use before the turn of the nineteenth century are difficult to identify, but records at Dartmouth allow us to list the following:

The works of James Ferguson in four volumes: Vol. 1, *Astronomy, explained upon Sir Isaac Newton's Principles;* Vol. 2, *Lectures on Select Subjects in mechanics, hydrostatics, hydraulics, pneumatics, and optics;* Vol. 3, *An Introduction to Electricity;* Vol. 4, *Select Mechanical Exercises: showing how to construct different Clocks, Orreries, and Sun-Dials.* These works were printed in London and went through many different editions,

two presidents of Dartmouth College comes from John Smith, because as a tutor more than as a professor he was responsible for the education of a particular class in all its aspects of learning during the first two years. Under the agreement he had made with Eleazar Wheelock, he was obviously not allowed to lecture as a professor on the subject of natural philosophy; yet this pathetic figure of a man, out of high purpose and conscientious devotion, felt the need, as tutor, of continuing with untiring industry to educate the class of 1780 in the field of natural philosophy during their junior year. In preparing for them a well-rounded educational program, he wrote a long series of letters to "My dear Class," outlining not only a full course in astronomy and the elements of geophysics, but referring his readers, also, to enough reference material in books on natural philosophy to provide a complete course in this subject.[4] He introduces the series as follows:

> According to my engagement, I shall, in the epistolary way, attempt something further for your entertainment and improvement. I am not induced to such an exercise to promote

---

most of which were printed by different publishing houses. "Miscellanea Curiosa, A Collection of Some of the Principle Phenomena in Nature." This was a scientific periodical publication aiming "to digest in a convenient Method, all the most curious Philosophical and Mathematical Discoveries, as they are to be met with, which may any way tend to the Use of Life, or Advancement of Arts and Sciences." It was basically written by Edmund Halley and is often referred to as Halley's Miscellany. Halley's name does not appear on the title page, however, and contemporary reference to it commonly attributed the work to W. Derham, whose name appears as the reviser and corrector after the first edition. Andrew Oliver, Jr., "Essay on Comets, in Two Parts . . . the whole interspersed with observations and reflections on the Sun and primary Planets." Published first in pamphlet form, it was subsequently printed as a small book: *Lectures on Comets, by Professor Winthrop, also, An Essay on Comets, by A. Oliver, Jr.,* published by W. Wells and T. B. Wait and Co. (Boston, 1811). Other books used were B. Martin, *Philosophia Britannica;* Wesley, *Philosophy;* and a *Dictionary of Arts and Sciences.*

4. The series of letters John Smith wrote to "My dear Class" are to be found in the Dartmouth College Archives. The introductory letter is undated; the one marked by Smith as "No. 1" is dated October 14, 1778, and they thereafter appeared weekly for the following ten weeks.

any sinister design. A desire to serve you, young gentlemen, is my only motive.

Whenever I reflect, that you are, with laudable and emulous industry, making such literary acquisitions, as, with divine grace, will qualify you for eminent utility to mankind, I feel an inexpressible pleasure; and think, I cannot be happier, than when I most labor to encourage and promote the noble pursuit.

Letter No. 1, dated 14th October 1778:

My dear Class,
Two years, you are sensible, have elapsed since you have resided in this seat of the Muses. The third has now commenced. —That the term of human life is measured by fleeting hours, we shall all experience, happily, if well employed; but sadly, if otherwise.
Your agreeable conversation, and application to study have inspired me with a belief, that your conduct has been influenced by a sense of *this*.—With the greatest pleasure have I hitherto conducted you in the paths of Science, and think it will not diminish, as we proceed in our journey.—It will be my constant endeavour to do everything, in my power, to serve you, and to this I am induced by sentiments of sincerest regard and affection.
I have thought a literary correspondence between me and you, might contribute a little to your advantages; to which I now invite you. And shall we find a topic, at present, for our purposes, more entertaining, or more instructive, than that of *natural Philosophy?* As that was a branch of your studies last year, and as it should be kept fresh in your minds, till the ideas are deeply impressed, I think it the most eligible.

About one half of the letters on natural philosophy which Professor Smith wrote to his class were on astronomy, and the astronomy book he was using and recommending was Volume I of James Ferguson's four volumes (note 3, above). In outlining the argument, which he continues over a number of the letters, he says: "To explain these phenomenon of the heavenly bodies, various systems have been invented; the most celebrated of which, are these four, the Ptolemaic, the Copernican, the Tychonic, and the New." About the "New" he writes:

You remember I promised to offer some arguments, for confirmation of the *New System.*—It has been observed, that mankind are almost always fond of the *new* and the *marvelous.* I trust, neither you, nor I shall be captivated with a principle, or system in science, merely on account of its novelty. No,— Truth must be the object of our invariable pursuit. Whenever we give our judgment, or opinion, we must, if we would act the part of rational beings, proportion our assent to the degree of evidence.

He then proceeds to follow Ferguson's adherence to *Sir Issac Newton's Principles* with a considerable amount of reference to failures of other systems. It is interesting to see the extent to which Professor Smith discussed the limits of validity of systems which he was not recommending to his students, particularly in a day when it was customary to present the students with the proper answer to be learned and not to allow discussion as an integral part of the educational process. As we might expect, however, he left no doubt in his students' minds as to what he believed to be correct: "I must give you a short account of one more, which is distinguished only by its extravagance, I mean the *Hutchinsonian.* Mr. Hutchinson, the author of *this,* calls Sir Isaac Newton and all his followers *senseless,* and *unphilosophical blockheads.*"[5] Smith outlines in detail Hutchin-

---

5. Professor Smith may not have really understood the conflict between Hutchinson and Newton. Although it is true that the astronomical explanations were very different, the philosophic dichotomy represented by the Newtonian and Hutchinsonian schools of thought was much more basically a conflict between the science and the religion of the day. The Hutchinson group was a militantly reactionary and religiously fundamentalist group that tried to prevent Newtonian ideas from disturbing the established notions of conservative theology. The most widely read book on this conflict was written by George Horne, Bishop of Norwich. The conflict between Newton and Hutchinson was a continuing one which culminated in the gradual elimination of the term "natural philosophy" in favor of the term "physics." Newton felt that reducing a problem to a mathematical equation was a real step in understanding nature, while Hutchinson and his followers considered this to be irrelevant. They argued that the causes of the motion of bodies were to be sought only in philosophy. There is no indication in the records that the Dartmouth undergraduates were made aware of the real issues in this argument.

son's theory: "I believe, by this time you think Mr. Hutchinson very romantic,[6] and that the hard names by which he called Sir Isaac Newton and his followers, may reverberate upon himself. The New, or (as it is usually termed) the Newtonian system, will triumph over every other."

The rest of the letters (there are twelve letters in all) have to do with the composition of air, the nature of density and pressure, and methods of measuring these physical quantities. Modern training in physics emphasizes the validity of significant figures and shorthand methods of representing large numbers. As is evident, however, from the trigonometry problem reproduced in the previous chapter, the 18th century mathematicians and natural philosophers did not embrace this concept. That point is illustrated by one short paragraph of Smith's on the calculation of the pressure of the atmosphere: "Now, since the earth's surface contains 200,000,000 square miles, and every square mile 27,878,400 square feet, there must be 5,575,680,000,000,000 square feet on the earth's surface, which multiplied by 2,160 pounds (the pressure on each square foot) gives 12,043,-468,800,000,000,000 pounds for the pressure or weight of the whole atmosphere."

In connection with all kinds of phenomena that can occur in the atmosphere, John Smith naturally gets into the problem of lightning, which he believed to be entirely chemical in nature and easily explained in the following manner:

> When the effluvia, which arises from acids and alkaline bodies, meet each other in the air, there will be a strong conflict or fermentation between them; which will sometimes be so great as to produce a fire . . . When you consider the effects of fermentation, you will not be at a loss to account for the dreadful effects of thunder and lightning: For the effluvia of sulphurous and nitrous bodies, and others, that rise in the atmosphere, will ferment each other, and take fire very often of themselves; sometimes by the assistance of the sun's heat.
>
> If the inflammable matter be thin and light, it will rise to the

---

6. The word "romantic" as used here by Smith is undoubtedly the archaic usage, meaning unreal, impractical, unrealistic.

upper part of the atmosphere, where it will flash without doing
any harm; but if it be dense, it will lie nearer the surface of the
earth, where taking fire, it will explode with a surprising force;
and by its heat rarify the air, kill men and cattle, split trees,
walls, and rocks, and be accompanied with terrible claps of
thunder.

The heat of lightning appears to be quite different from that
of other fires; for it has been known to run through wood,
leather, clothes, & c. without burning them; while it has broken
iron, steel, silver, gold and other hard bodies. Thus it has
melted, or burnt asunder a sword, without hurting the scab-
bard, and money in a man's pocket, without hurting his clothes;
The reason of this seems to be, that the particles of the fire are
so fine, that they pass thro' soft, loose bodies, without dissolving them; while they spend their whole force upon the hard
ones . . . When it thunders and lightens, it commonly rains too,
the same shock driving together, and condensing the clouds—
And this is the reason why it often rains so violently in a thun-
derstorm.

It is interesting to speculate in this connection on the un-
derlying reasons why the Rev. John Smith chose to present his
class with only the chemical theory of lightning, particularly in
view of his lectures on celestial mechanics in which a number
of conflicting theories were discussed in some detail. Benjamin
Franklin's *Experiments and Observations on Electricity* had
been published in 1769, and these were widely read throughout
the intellectual world.[7] Experiments using an electrical machine
were described in detail in Ferguson's *Electricity,* which Mr.
Smith undoubtedly had at his disposal. Not many miles south,
in Woburn, Massachusetts, Loammi Baldwin (the discoverer of
the famous New England apple by that name) had made himself
locally famous some years before by repeating Franklin's electrical kite experiment and lighting himself up "like a glow-

---

7. It was through Joseph Priestley's *History and Present State of Electricity* (London, 1767, 2nd ed 1769—the one usually cited) that Franklin's
experiments became widely known. Franklin gave to Priestley the precise
details and read the manuscript, although Franklin's original description
of his kite experiment had appeared many years before in Royal Society,
Philosophical Transactions, Vol. 48 (1753), p. 567.

6. George Ticknor's fat little professor.

worm," much to the terror and consternation of his family and friends who witnessed the event.[8] One's suspicions are considerably aroused that Professor Smith's attention to classical scholarship as well as Professor Woodward's administrative responsibilities gave Dartmouth College at this time no one who was actually keeping abreast of developments in the field of natural philosophy.

Although a very slender man when he was young, Professor Smith, as he grew older, put on considerable weight. By the time he was one of the elders of the town, he was by far the fattest man in the vicinity. As far as can be determined, no formal portrait was ever painted. In July of 1803, however, an eleven-year-old member of the Class of 1807, George Ticknor, painted a "view of the principal buildings of Dartmouth University." In the foreground of the Ticknor picture is the representation of a very rotund little figure, and this has been presumed to be a likeness of John Smith.

In contrast to Bezaleel Woodward, John Smith was a failure as an 18th-century college professor. He was shy, absent-minded, and bookish, and he was taken advantage of by students and faculty alike. His scholarship would have assured him an honored place in a modern university, but not at Dartmouth in the late 1700's. It is a revealing commentary on the state of New England higher education to realize that two hundred years ago it was more important for the students to learn by rote than to be encouraged to think independently. It was more important for the faculty to reproduce slavishly year after year exactly the same material than to innovate by writing long letters on Natural Philosophy to "My dear class" or to write grammars in Latin, Greek, Hebrew, and Chaldaic.

---

8. A brief description of Loammi Baldwin's experiment with lightning is given on p. 7 of Sanborn C. Brown, *Count Rumford, Physicist Extraordinary* (Garden City, N.Y., Anchor Books), 178 pp.

Chapter 3

NATURAL PHILOSOPHY DRIFTS

IT should be evident by now that in the early days of Dartmouth College the subject of Natural Philosophy was not considered important enough to occupy the full attention and energies of any single man on the faculty. Just as Bezaleel Woodward and John Smith spent the bulk of their time and effort in other areas, it is not surprising to find that on Woodward's death the professorship of mathematics and natural philosophy went to a man who had, to be sure, already earned a reputation in the educational field but whose contribution in natural philosophy was so insignificant that no record of it now exists.

John Hubbard, a member of the Class of 1785, was called to Dartmouth from Deerfield Academy in Massachusetts, where he had been headmaster from 1802 to 1804. He had distinguished himself as an educator by publishing two books. *The Rudiments of Geography,* written in 1803, was dedicated to "The Honourable John Wheelock, L.L.D., President of Dartmouth College." *The American Reader,* published in 1804, was used in schools to teach the art of oral reading, and was reprinted in many editions.

Hubbard's real interest was in music. His position at Dartmouth allowed him to expand his horizons in this area, although he already had local reputation as a composer.[1] He published

---

1. In reporting the twentieth anniversary celebration of Dartmouth's

an essay on music, given before the Middlesex Musical Society, September 9, 1807, at Dunstable, Massachusetts. In it he extols the pure musical forms of Handel rather than the then currently popular experimentation with the fugue style, which emphasized form rather than expressing emotion through music. After his death in 1810, his wife found and published a manuscript on sacred music containing "Thirty Anthems selected from the works of Handel, Purcel, Croft, and others" which Hubbard had collected and arranged during his lifetime.[2]

The appointment of John Hubbard, a secondary-school educator and musicologist, to the professorship of mathematics and natural philosophy may seem strange to us today, but in 1804 it was quite appropriate. A rigid curriculum was still required of all students, and the faculty continued to be tied down completely by exacting schedules which were as inflexible as they were grueling. Most of the graduates of the College were planning to be ministers, and the interaction of science and religion was one of Professor Hubbard's principal themes. As the Reverend Elisha Parish said in his eulogy on Hubbard:

When as a philosopher, surrounded with the apparatus of science, extending his researches to the phenomena of the universe, amazed at the minuteness of some objects, astonished at the magnitude and magnificence of others, his mind was transported; when he explored the heavens, and saw worlds balancing worlds, and other suns enlightening other systems, his senses were ravished with the wisdom, the power, the goodness of the Almighty Architect. On these subjects he often declaimed, with the learning of an astronomer, the simplicity of an apostle, the eloquence of a prophet. He illustrated the moral and religious improvements of the sciences; the views of his students were enlarged; the sciences became brilliant stars to irradiate the hemisphere of christianity. The perfect agreement between

---

first commencement, the *Concord* (New Hampshire) *Herald* for August 31, 1791, reports that an "Ode to commencement" was performed, set to music composed by John Hubbard, "a man of superior talent, knowledge and taste, in the science of music."

2. Mr. Hubbard's volume of sacred music entitled "Thirty Anthems" was published in 1814 by E. Little and Co. at Newburyport, Massachusetts.

sound learning and true religion was a favorite theme of his heart.[3]

Although our knowledge of the details of Hubbard's instruction in physics is not great, from a newspaper account we do know something of the curriculum in 1805 and can read "that the Juniors shall recite Adams' Lectures on Natural Philosophy, Smellie's Philosophy of Natural History, Ferguson's Astronomy, Harris' Hermes, Paley's Moral Philosophy, Horace, Composition and Oratory."[4]

Adams' work covered, in four volumes, the whole of classical physics,[5] and Ferguson's thick book on astronomy delved much further into the field of celestial mechanics than any modern undergraduate would be expected to go. If it is literally true that the juniors at Dartmouth College had to absorb so well all the contents of these five volumes that they could recite it comprehensively on call, the amount of physics which every member of the College at that time had to know before he was graduated far exceeds our present requirement.

As in the case of Professor John Smith, the only student record we have of Professor Hubbard's teaching is in mathe-

---

3. Eighteenth and early nineteenth century eulogies are almost never useful historical or biographical sources. Occasionally, however, they do shine some light on certain facets of personality, and Elisha Parish's eulogy on John Hubbard is helpful in this regard. It was pronounced at Dartmouth College in September 1810 and was printed in that year by C. W. S. & H. Spear in Hanover.

4. Our information on the 1806 curriculum comes from the September 27, 1805, number of the *Dartmouth Gazette*. As printed in this source, the quoted passage reads, "Smellie's History of Natural Philosophy." The only recorded work close to this title is "The Philosophy of Natural History" by William Smellie (1740–1795). We are assuming that the *Dartmouth Gazette* made an error.

5. George Adams, who wrote *Lectures on Natural and Experimental Philosophy*, was "Mathematical Instrument Maker for His Majesty, &c." There was an American edition published in Philadelphia in 1807. As has been noted, the term "physics" was not used in the eighteenth century. The word began to gain popularity about the turn of the century. In the appendix to the fourth volume of this work we find its early use: "a brief outline of Physics, a Natural Philosophy, in the form of collegiate Examination."

matics. In the Dartmouth College archives is a carefully prepared manuscript by Samuel A. Kimball, a graduate of the class of 1806, whose formal copying of "A System of Spheric Trigonometry by John Hubbard, Esq., Prof. Math. and Nat. Phil. at Dartmouth University, April 15, 1805," provides us with details of the teaching in this subject. A rather interesting side note comes from one of the home problems under "Miscellany" in this notebook: *"To measure the height of the Atmosphere.* The Sun's rays being refracted by the atmosphere cause him to give light when 18 degrees *below the horizon.*" As we read in the accompanying illustration, the answer reached is "the height of the Atmosphere [is] 45 miles." In Mr. Hubbard's previously mentioned *Rudiments of Geography* there is included "an Introduction explaining the Astronomical parts of Geography to which is added a chronological Table of the most important events which have happened from the Creation of the World to the present day." In this introduction there is a paragraph entitled "Atmosphere," where the same atmospheric height is featured: "The earth is every where surrounded with air, or an atmosphere, extending about 45 miles from its surface. This atmosphere affords us breath, conveys sounds, supports the vapours, and answers many other useful and necessary purposes."

Professor Hubbard's scientific leanings are also evident in his choice of "the most important events which have happened from the creation of the world to the present day." We find entries in his *Geography,* such as:

1492　America discovered by Christopher Columbus
1494　Algebra first known in Europe
. . . .
1602　Decimal Arithmetic invented at Bruges.
1608　Galileo, an Italian, first discovers the satellite of Saturn with a telescope, then just invented in Holland.
1614　Logarithms invented by Napier of Scotland.
1616　The first permanent settlement made in Virginia.
1619　Dr. William Harvey discovers the circulation of the blood
1620　New England settled.
1626　The Barometer invented by Toricelli.

Miscellany. To measure the height of the Atmosphere. The Sun's rays being refracted by the atmosphere cause him to give light when 18° below the horizon. — Let H.O. be the sensible Horizon. C.A a ray of light passing from the Sun. Suppose a spectator at B, he will see the Sun 18° below the horizon. The diameter from C to D, is the height of the atmosphere.

By Trigonometry — — Say,
As the ∠ of angle at C = 81° — — 9,99462
to the side B.A = 4000 — 3,60206
So is radius, 90° — — 10
to the side C.A = 4045 — 13,60206 — 9,99462 = 3,60744 = ...

Subtract A.B. from A.C. or 4000 from 4045 and we have C.D. the height of the Atmosphere 45 miles —

H C   B .... O

S

The Earth.

7. A Dartmouth junior's problem in spherical trigonometry in the early nineteenth century.

1627 The Thermometer invented by Drubellius.
1649 Charles I. beheaded at Whitehall.

· ·

1783 Peace ratified between Great Britain and the United States.
1783 The memory of Handel commemorated at Westminster Abbey, by about 500 musicians.
1789 Revolution in France begins; destruction of the Bastile

It is significant that Professor Hubbard interspersed scientific events with historical ones. It is also completely in character that the commemoration of Handel stands with the peace ratified between Great Britain and the United States as the other important world event in that year. To quote again from Elisha Parish's eulogy:

Mr. Hubbard's industry, and native energy of mind made him a distinguished proficient in the abstruse sciences. From early life he was delighted with the poets, and could himself "build the lofty rhime." Music, the kindred art, ruled his affections to the last moment of his life. . . . In this divine art his attainments were equalled by few persons in our country. Animated with pure devotion, enlightened with just views of public worship, conscious of the irresistable power of musical sounds, to move the passions, and produce corresponding sentiments in the heart, he ardently engaged to promote a just style of sacred music. The gay and volatile airs which have been so long in vogue, he believed to be fatal gales to dissipate the serious thought, and devout affections, produced by other branches of worship . . . At the head of a musical Society, extensive in his influence, and highly respected by all lovers of sacred song, had his life been spared, he would probably have done much to promote christianity, by improving the music of our churches. He had a happy facility in illustrating the practical advantages of every science. He not only explained its principles; but traced its relation to other branches of knowledge. Not satisfied by merely ascertaining facts, he explored the cause, the means, the ultimate design of their existence.

The musical Society mentioned here is the Handel Society at Dartmouth in the founding of which in 1807 John Hubbard played a major role. For some years the society met at Professor

8. Portrait of Professor John Hubbard owned by Dartmouth College.

Hubbard's home on Friday evenings for the discussion of musical theory, as well as for singing and playing. A cellist and a composer, he was also a critic and a philosopher of music. To quote from his "Essay on Music":

"In treating upon music, we shall consider it both as an art and a science. As an art, it depends upon the powers, abilities, and genius of the writer. As an art, it cannot be limited, or restricted with any particular rules. The genius, the feelings, and the improved taste of mankind, must regulate every good writer. Like the painter, the sculptor, the architect, and the poet, nature and propriety must direct the effusions of his mind. As a science, it is regulated by measure, harmony, cadence, accent, mode, & c. Science may invent good harmony, agreeable measure, flowing and easy cadence; but genius only can give force and energy to music.

"We shall consider the essential parts or divisions of music, as consisting of *melody, harmony, expression,* and *accent.*"

It is easy to see from this essay why it was that Hubbard was greatly concerned about good oral reading, and in his introduction to *The American Reader* he writes:

No attention in forming the proper modulations in tone to the voice can be too great. The whole design of reading aloud is to convey information to others. The reader then, like the speaker, should use every effort to make himself fully and plainly understood. This he never can do, unless he fully comprehend the subject he is reading, and use the same modulation of the voice, tones and emphasis, which a good speaker would.

The principle objects of attention in reading and speaking are Pronunciation, Emphasis, Modulation of the voice, and Cadence.

Professor Hubbard's death on August 14, 1810, was sudden and unexpected and was considered a tremendous loss to the College. His family was left in very desperate circumstances financially, and the junior class "presented a mourning suit to the widow which she received with uncontrollable emotion." The inscribed marble slab which covers his grave in the Hanover cemetery was also a gift, from the undergraduate organization known as "The Social Friends."

Chapter 4

ADMINISTRATIVE TURMOIL

THE evolutionary development of an educational institution is an intricate interplay between the social structure of the society in which it is immersed, the aspirations of its students, faculty, and administration, and to a remarkable degree the individual personalities of men in leadership roles in a college community.

For the first sixty-five years of Dartmouth College's existence, the place of natural philosophy in the undergraduate curriculum was as static as its offerings were rigid. The last of the figures who existed on this educational plateau was Ebenezer Adams, Professor of Mathematics and Natural Philosophy from 1810 to 1833. Professor Adams' training was much like that of his predecessors. He was graduated from Dartmouth in 1791 and for the following fourteen years was preceptor of Leicester Academy. In 1806 he went to Portland, Maine, to teach at Portland Academy, and on the death of Professor Hubbard in 1810 he came to Dartmouth and to the Professorship of Mathematics, Natural Philosophy, and Astronomy, a position for which he had been considered when Bezaleel Woodward died in 1804. At that time John Hubbard was the first choice of the trustees. Since it was necessary that the chair be filled immediately, Ebenezer Adams had been designated as an alternate in the event Hubbard declined. As has been seen, however, John Hubbard accepted the appointment in 1805.

Ebenezer Adams was qualified in at least one discipline of his new professorship. Professor Woodward had introduced the standard arithmetic text by Nicolas Pike, which was published in 1788. The second edition of 1797, "enlarged," had been "Revised and Corrected, by Ebenezer Adams, A.M. Preceptor of Leicester Academy."[1] We can be sure President John Wheelock knew of this when he invited Adams to Dartmouth College in the first place. After his appointment, however, Adams did no further work on mathematical texts, for he became more and more involved with administrative matters along with his teaching duties.

There is one anecdote that has been told and retold about Professor Adams, who was known for a ponderous and pedantic method of teaching rather than for any quickness of wit. Obviously the incident proved remarkable because it was so surprising: "In illustrating astronomy by the aid of a broken old orrery, the misbehavior of the moon drew from him the remark that she seemed somewhat lunatic. The joke was so good, besides being his sole production, that he would never suffer the machine to be repaired (it is still broken) and thus he preserved the joke for each succeeding class." Figure 9 shows an old orrery from the College Museum which certainly could have been the very object of this tale.

Under Professor Adams' tutelage the first original scientific observations were begun at Dartmouth, when his fourteen-year-old son, Ebenezer, Jr., started keeping a "thermometrical regis-

---

1. The second edition of Pike's book was published by a different publisher from the first, which had been issued by Isaiah Thomas of Worcester, Massachusetts, in 1797. In the publisher's preface to the new edition is found this statement: "Much attention has been given to the Revision and Correction of this valuable work, at the request of the Proprietor, and recommendation of the Author, by the ingenious Mr. Ebenezer Adams, Preceptor of Leicester Academy; and, as far as his other engagements would permit, the Author himself has aided in making the work as perfect as possible." A hint as to why the "Author himself" did not work on editions following the first, is found in the preface to the third edition, which was "Revised, Corrected, and Improved" by Nathaniel Lord and published by Thomas and Andrews at Boston in 1808: "Application was made to the Author, requesting him to revise and improve the work for a new Edition; but he declined on account of want of health."

9. Dartmouth's 18th century orrery.

ter." This was the beginning of the meteorological observations which have been recorded by the College ever since. Although Adams performed his teaching duties conscientiously and adequately, his real contribution to Dartmouth's welfare was administrative, and on more than one occasion, the full weight of the college's problems landed on his shoulders as acting president.

Dartmouth was slow to break away from the rigidity of 18th-century higher education and to accept not only new discoveries in natural philosophy but also the growth of a whole new concept of education. The reason for this can be laid at the door of President Wheelock, although he was only a product of the era in which he lived. The history of New England colleges and academics is the history of individual men with a missionary zeal and singleness of purpose who dedicated their lives to propagating the education which they themselves had received. In the 18th century these were not considered public institutions, but private ventures with many similarities to a family business. Eleazar Wheelock ran his college with the autonomy of a dictator, and he brought up his son John to follow in his footsteps. John, however, was not happy with this business and strains developed. There was no doubt in Eleazar Wheelock's mind that Dartmouth was a Wheelock enterprise and should continue to be a property of the family. Of the three first professors, Ripley, Woodward, and Smith, only Smith had not married a Wheelock daughter, but, as mentioned before, Smith was notorious for his complete subjugation to authority.

Most of the troubles of John Wheelock, the second president, were probably beyond his own control. When he became president of Dartmouth in 1779, apparently without really deciding he wanted the position, the family-ownership concept of colleges was beginning to be displaced by the concept that higher education was in the public domain. Also, he had a personality which antagonized people, and when he got into battles with his trustees, with the faculty, and with the church, he never seemed to understand that he could achieve more by standing up to his adversaries face to face than by maneuvering behind their backs. It was in the turbulent era of John Wheelock's ad-

ministration and its aftermath that Ebenezer Adams was called
upon to carry much of the responsibility for the College's very
existence, and it was the conflicts of these years that kept educational progress at Dartmouth so long at a standstill.

The histories of Dartmouth chronicle the strife of those
years in much more detail than is necessary here.[2] On the other
hand, one cannot understand the development of any part of the
College without some brief review of the ordeal that the College
and all its officers went through in the bitter years between 1804
and the decision of the Dartmouth College Case in 1819. To
quote Richardson's analysis of this stormy period:

> Despite all the ramifications of the contest, despite all attempts
> put forth on both sides to confuse the issue by attaching it to
> fundamental principles advanced by political party or religious
> sect, there was in reality throughout the whole controversy but
> one point which the contending parties deemed of importance.
> Namely, whether the control of the College should rest with
> the Wheelock dynasty or with the trustees. While the matter
> became involved in partisan disputes as the quarrel progressed,
> so that for a time political campaigns centered around the con-
> troversy, while the theological champions of the time found in
> the claims of the two factions effective material with which
> their own dissensions might be embittered, and while, at least,
> constitutional issues of the highest importance rose from the
> questions involved, nevertheless these were but extraneous is-
> sues—eagerly seized by one or the other of the contending par-
> ties and of interest to them only as contributing to the solution
> of the one question which, in their minds, was really vital—
> should John Wheelock or the trustees rule?

The squabble began in a way that was almost peripheral
to the College. In 1804 the trustees appointed the Reverend
Roswell Shurtleff, then a tutor at the College, to the professorship of theology. It had been the general assumption that the

---

2. Professor L. B. Richardson has written a most readable two-volume
History of Dartmouth College which was published in Hanover by Dartmouth College Publications in 1932. His two chapters appropriately called
"Storm and Stress" and "Depression and Recovery" give, in about a hundred pages, the general details of this controversy.

professor of theology would serve as pastor of the College church. As mentioned before, Professor John Smith had been acting pastor of the Church of Christ at Dartmouth College; this was interpreted by all concerned as on a temporary basis, until a theological professor could be appointed. Thus, when Shurtleff was named to this chair, the church sent Shurtleff a regular call to its pastorate, and the members were undoubtedly delighted after twenty years to have the prospect of a respite from the monumental dullness of John Smith. President Wheelock had a completely different idea, however, and announced that Professor Smith's pastorate was by no means finished. As a compromise he suggested a co-pastorate for the two parishes of the church, one across the Connecticut River in Hartford, Vermont, and the other, in Hanover, New Hampshire. The church organization, as originally constituted, included inhabitants in the surrounding country and, although the people of Hartford had built their own church, Wheelock had previously persuaded them to rely completely on college tutors for their services, since they could be obtained at a "cheap rate." At the meeting to discuss his scheme of a co-pastorate for Hartford and Hanover, Wheelock made sure the inhabitants of Hartford were present in force. They produced a voting majority for Wheelock, and co-pastors for the two churches were forced on the unwilling Hanover members, who were infuriated by this action.

As time went on, more and more issues arising out of this quarrel in the local church grew to larger and larger proportions, until a bitter fight was imminent between Wheelock and his Board of Trustees. The open break between the President and the Trustees came in November of 1814 when the Trustees passed a vote which read in part, "In order to relieve the president from some portion of the burdens which unavoidably devolve upon him, in the future he be excused from hearing the recitations of the senior class in Locke, Edwards, and Stewart." This was in fact a move to strip Wheelock of his responsibility of teaching the senior class, which he himself had done for 35 years and which his father had done before him throughout the decade he was president. John Wheelock did not take this lying down. He wrote and published a scathing pamphlet against the

trustees, sent copies to all the state legislators, and moved to
Concord to lobby for the state to take over the College. In August of 1815 the Board of Trustees removed John Wheelock
from his position as President and appointed in his place Francis Brown, a graduate of the Class of 1805 and tutor from 1806
to 1809, who had become a minister in North Yarmouth, Maine,
after refusing Wheelock's offer of John Hubbard's professorship
of languages in 1810.

Under John Wheelock's urging, the state legislature passed
a bill which transformed Dartmouth College into Dartmouth
University and provided state control over the institution. From
here on, the controversy became a legal one, centering upon the
right of a state's legislature to alter the terms of a charter given
to a private institution by an English King prior to the Revolution. The conflict was finally resolved by the Supreme Court
of the United States on February 2, 1819, in favor of Dartmouth
College.

These chronological facts in no way convey the turmoil and
stress which pitted two faculties, two student bodies, and two
congregations against one another. The University which controlled all of the College buildings had a complete faculty but
a very small number of students. The College, on the other
hand, occupied makeshift quarters and had a complete faculty
(although the state legislature had set a legal fine of $500 against
anyone who should teach in the College), and nearly all the students remained loyal to it. John Wheelock became president of
the University; William Allen, his son-in-law, held the chair of
theology and ultimately succeeded to the presidency; and a
James Dean was appointed the University's Professor of Mathematics and Natural Philosophy. The College faculty consisted
of President Brown, Professor Ebenezer Adams, and Professor
Roswell Shurtleff as Professor of Theology. Professor Adams
was the ranking member of the faculty and therefore served as
Acting President whenever Francis Brown was absent. At the
height of the controversy this was very often, since the President
was required to go to Concord, Boston, and wherever legal, financial, and political help could be secured. Francis Brown had
come to Hanover a strong and vigorous leader, willing to aban-

10. Professor Ebenezer Adams at the age of 66. The portrait hangs in the Baker Library at Dartmouth.

don his career as a minister to help the College. A few months
after his cause was won, he collapsed, exhausted, and died a year
later on July 27, 1820, in his 37th year, "the last and most costly
sacrifice in the struggle which had ranged so long."

On President Brown's death Professor Adams again became Acting President. Within a month, however, the Reverend
Daniel Dana was elected President, but he found college administration was far from his talents, was absent from Hanover most
of the time to restore his health, and finally resigned after less
than a year in office. The Trustees next asked the Reverend
Gardner Spring of New York to serve, but he declined after a
long period of indecision. Finally, in February of 1822, the
Reverend Bennett Tyler became President of Dartmouth College. Thus, from the summer of 1819, when President Brown's
physical collapse made it impossible for him to act effectively,
until February of 1822, the duties of the presidency had fallen
to Ebenezer Adams.

By the time Adams was relieved of the administrative responsibilities which had always been added to his duties as Professor of Mathematics and Natural Philosophy, he was 57 years
old. It is not surprising that during the next eleven years before
his retirement from the faculty, he made no important changes
—either in the curriculum or in his teaching approach. He was
no innovator; he was a holder of the line through thick and thin,
and the College owed him a tremendous debt when he died in
Hanover, July 15, 1841. At his death, the undergraduate publication, *The Dartmouth,* commented:

> Among the pupils of Professor Adams, many of the living gen-
> eration of accurate thinkers and true men are found in every
> department of life. And it is but just to say that no teacher in
> New England has been more successful in forming the minds
> of the young to precision and distinctness . . . He had no misty
> conception of truths and he could not tolerate inaccuracy . . .
> During his connection with the College, it attained the reputa-
> tion it has so long enjoyed of making sound, able, working men.
> To this enviable good name, no man has contributed more.
> The intellectual qualities of Professor Adams were of the high-
> est order. He was, by nature, equal to any effort. His literary

attainments were, from taste and the nature of his duties as a daily teacher, a good deal confined to a single department, but his spirit was catholic and his opinions in literature, morals, politics, and religion always clear and philanthropic.[3]

3. *The Dartmouth's* biographical sketch of Ebenezer Adams was published in Vol. 3, issued in September 1841, and is found starting on p. 33.

Chapter 5

SCIENCE CAN BE EXCITING

A major requisite for good teaching today is a vital interest, and a modicum of activity in, original research or other specialized scholarly pursuit within the teacher's own discipline. If teachers remain too isolated, both intellectually and geographically, and are too laden down with routine and overly heavy teaching schedules, neither the teachers nor their institution prospers. The early history of natural philosophy at Dartmouth College is a classic illustration of this fundamental principle in operation.

In the organizationally minded society of the present day, most colleges and universities, as well as national professional societies, provide for sabbatical leaves, visiting and exchange professorships, national lecture tours, traveling exhibits, and reasonable teaching loads, for the benefit of both the teacher and his teaching. But this is the result of an educational evolution consciously fostered to avoid the intellectual isolation of the Woodwards, the Smiths, the Hubbards, and the Adamses.

Progress has been achieved over the years by the energy and enthusiasm of individual professors, and changes in the teaching of natural philosophy at Dartmouth College were not the result of administrative outlook or executive fiat. Indeed, Nathan Lord, who came to the presidency during Ebenezer Adams' professorship and who held office for the next thirty-five years, said in his inaugural address that he disapproved of the "restless age in which we live and the new inventions which are changing so rapidly the civilization of our fathers." He announced that he planned to turn his back on the "crowds of innovations in education" that were then being proposed to take the place of the time honored and tested methods of old. Although he recognized the advantages of trying novel ideas, he felt that Dartmouth should hold fast to traditional methods until there was overwhelming evidence that the change would in fact be advantageous. He reckoned, however, without Ira Young.

Ira Young was the first of the moderns. Woodward was tutor, treasurer, trustee, vice-president of the college, first librarian, professor of mathematics, and, finally, professor of natural philosophy. Smith was a tutor and professor of learned languages and never held the rank of Professor of Natural Philosophy. Before Hubbard turned to science he was a geographer and a musicologist. And Adams was a professor of languages prior to his designation as a professor of natural philosophy. But Young, although trained in the classical tradition and despite the fact that he had served as a tutor for three years, was a scientist from beginning to end, a specialist in mathematics and natural philosophy, a leader among the educational innovators President Lord so mistrusted.

The man who was to lead science teaching out of the doldrums at Dartmouth College was initially trained as a carpenter in Lebanon, New Hampshire. He wanted desperately to have an academic education, but his father was determined to keep him in the family trade. Until the age of 21 he worked for his father, although every winter after he was 16 he taught in one of the district schools in the neighborhood. President Lord says of him "He had been occasionally sent by his father to work as a carpenter, in these halls and libraries. He had, insensibly, caught the spirit of the place . . . The Professor [Ebenezer Adams] had given him leave, at his recesses of work, to inspect the apparatus. He experimented with it, took it in pieces, scrutinized, repaired and cleaned it, of his own notion, to the delight of the Professor."[1]

_____

1. President Nathan Lord's "Discourse Commemorative of Ira Young," was published by the Dartmouth Press, Hanover, April 1859.

When finally free of his indenture to his father at 21, he
went to Meriden Academy, where after one year of study he was
deemed prepared for college. He was graduated with the Dartmouth Class of 1828, taught in various schools for two years, and
was then called to Dartmouth as a tutor in 1830. His outstanding ability and vigor were quickly recognized, and when in 1833
Prof. Ebenezer Adams asked the trustees to relieve him of his
duties on account of his age, he not only recommended that Ira
Young be appointed Professor of Mathematics and Natural Philosophy, with which the trustees agreed, but moreover gave his
youngest daughter, Eliza, in marriage to his brilliant protégé.

It is conceivable that the trustees were worried about Ira
Young's ability to fill the role of a professor. Every previous professor of mathematics and natural philosophy had proven his
scholarship by writing or editing secondary-school mathematics
texts. Young came with no such qualifications, nor even with a
particular background of distinguished classical scholarship.
They soon found, however, that they had in him a professor
whose whole energies were directed toward putting the instructional program in astronomy and physics into first-class form.
Within a very short time of accepting his new position, Professor Young informed the trustees that the few pieces of apparatus
which the College owned were in terrible condition and that
actually the scientific equipment had never been of proper
quantity or quality compared with other institutions. Dartmouth, he advised, was falling behind, compared with sister colleges which were busily engaged in upgrading science instruction. The force of his arguments made its impression on the
College's board of trustees, and they not only gave him a reasonable amount of money to purchase new equipment, but also
authorized expenses for him to travel around New England in
order that he might become familiar with what was going on in
other colleges.[2]

---

2. The Trustees action to upgrade the College's philosophical appa-
ratus followed a "Report of Faculty on Library and Apparatus, 1846."
The report is in the form of a letter preserved in the Dartmouth College
Archives, dated July 27, 1846, written in the hand of Charles B. Haddock.
The other member of the committee was Ira Young. The request to the

Five years after Young became a professor, the Trustees voted to construct the building known as Reed Hall, and because of Young's previous experience in the building trade, he was asked to represent the College in obtaining architectural plans and in arranging for a contractor. Although the Trustees' minutes somewhat pointedly suggest that care needed to be exercised to make sure the College was not somehow victimized, Reed Hall was constructed by Young, Young, and Young: the architect was Ammi B. Young, the builder was Dyer B. Young, and the College's interests were superintended by Ira Young—all brothers from Lebanon, New Hampshire. (Professor Young was, as might be expected, careful to plan space in the new buildings for housing philosophical apparatus and for various other scientific collections.)

Interest in natural philosophy was so stimulated by the new philosophical apparatus, the new accommodations in Reed Hall, and Professor Young's enthusiasm that when the trustees decided in 1841 that another appeal for public support should be made, the largest donation endowed a professorship in natural philosophy. Samuel Appleton of Boston gave $10,000 for this purpose, and on his death in 1854 the College received an additional $15,000 to maintain and improve the Department of Natural Philosophy.

Young never seemed to stop badgering the trustees, and they certainly responded generously to his requests. Perusal of the Trustees' minutes reveals no entry in which Young's requests were either turned down or even pared. In 1838 Young became Professor of Astronomy in addition to holding his Professorship of Natural Philosophy. This was a manifestation of a growing interest on his part—an expensive one, as it proved, for the College. In 1846 he was granted $2,300 to buy a six-and-one-half-foot refracting telescope from Germany, which was delivered in this country in May of 1848. The import duty was 30 percent, and this unexpected drain on college funds was such a

---

faculty committee to increase the holdings of library books was not recommended, although Charles B. Haddock was the College Librarian, but the recommendation to spend money on philosophical apparatus was enthusiastically endorsed.

11. Early portrait of Professor Ira Young, belonging to his great-granddaughter Sara Proctor Fay and now hanging in her home in Cambridge.

blow that Dartmouth petitioned Congress for a release. A special act was passed to eliminate the tariff, and the telescope finally reached Hanover in September of that year.

Typical of Young's mode of operation, the telescope was obtained but with no place whatever to put it. He applied again to the Trustees and was awarded the sum of $250 to build a small observatory in his garden. The little building, 28' × 13', was divided into two rooms. One compartment had a sliding roof and was furnished with an old millstone for a pier on which to mount the telescope. The second room had a plastered ceiling and was equipped with a stove and "other conveniences for a computing room." This center of scientific activity aroused a tremendous interest among the students. The instruments were used almost constantly whenever the weather was good, either for instructing the students or showing visitors, or for regular observations by Professor Young himself. Young wrote to the trustees: "I trust the interest thus awakened among the students will have its appropriate reflex influence upon the Honorable Board in stimulating them to use all proper means of early completion of the promised observatory and instruments." One might call these high-pressure tactics, but they certainly paid off for Professor Young's objectives. Usually the minutes of the meetings of the Board of Trustees were drafted with two or three lines per item of business, short and to the point, but for the special meeting of December 1852, this precedent was completely shattered by the insertion of almost eleven pages of correspondence.

Reverend N. Lord, D.D. Pres., Dart. Coll.
Dear Sir:

On the 21st Oct. last Dr. Geo. C. Shattuck of this city put into my hands his check for $7000 in special trust, for the purpose of enabling the Trustees of Dart. College to erect an Observatory near the College and supply it with instruments to illustrate the principles of Astronomy, Natural Philosophy, and the other natural Sciences.

From the full documentation on the Shattuck gift, faithfully copied in the Trustees' Minutes, it is known that the donor's particular request included these points:

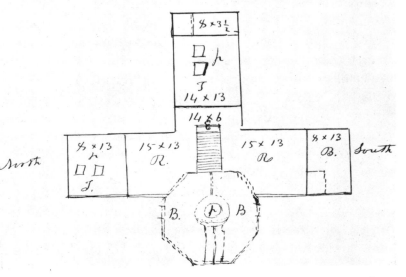

12. A reproduction of the sketch of Ira Young's observatory, from the Trustees' Records. The pier for the telescope is an old millstone from Professor Young's garden. Professor of Astronomy John M. Poor, writing about the changes to the Observatory made in 1908 commented: "Examination of the original brick pier built in 1853 showed that it was entirely inadequate. It proved to be rather cheaply built for the most part consisting of air upon which rested a stone which had previously seen service as a mill-stone" (undated ms in Dartmouth College Archives).

1. The Observatory should be built of hard-burnt brick and the walls coated outside with mastic cement. The dome as well as the roof should be covered with metal and a suitable cistern or tank should be provided for the collection of Rain Water for the use and security of the Building: A.B. Young, Esq. to be the Architect and Prof. Young to be the superintendent for building the Observatory.
2. The Trustees should authorize Prof. Young to go to Europe for the purchase of the Apparatus mentioned in his letters; and for the purchase of such Books, as could be better purchased abroad.

Dr. Shattuck warned that his gift would not be forthcoming unless the trustees added to his $7000 another $4000. However, at the meeting in which this matter was discussed the trustees voted to produce the $4000, borrowing it if necessary.

Professor Ira Young was by no means moderating his interest in physics while working so hard to initiate a real department of astronomy. It is of significant historic interest to reproduce his thinking on the direction of growth in which his teaching in natural philosophy was to lead him. On October 5, 1852, he wrote to Dr. Shattuck:

My Dear Sir:

In accordance with your request I will now proceed to point out what Apparatus is needed in my department, to render our instruction most efficient and useful to our pupils.

In *Mechanics,* we absolutely need a *rotascope* (an instrument for illustrating the principles of rotary motions) and apparatus for illustrating the Laws of *percussion,* and also for demonstrating experimentally the *Laws of friction.* These would probably cost about sixty dollars.

In Hydrostatics and Hydrodynamics, we need for the purchase of articles for demonstrating and illustrating the Laws of liquids, at rest and in motion, an outlay of about forty dollars.

A good hydrostatic balance with its accessories for determining the exact *specific gravity* of substances, is a very important instrument for scientific investigations and is owned by all of our more important Colleges, but it is expensive costing about $125 and is not *indispensable* in the teaching of students.

In *Pneumatics* we very much need a *condensing fountain*

and *apparatus* for demonstrating the Laws of pressure and the elasticity of gases. These would probably cost about fifty dollars.

In *Acoustics,* we are entirely deficient of apparatus. On account of its expense, I have hitherto deferred purchasing a complete set for illustrating the laws of sound, and their application in the construction of musical instruments. Costs in Paris about one hundred and fifty dollars.

In *frictional Electricity* the addition of articles costing fifty dollars would make our demonstrative apparatus tolerably complete.

But in *Magneto-Electricity* and *Electro-Magnetism* we need an outlay of nearly one hundred dollars to enable us properly to demonstrate the Laws, and applications of this new and deservedly popular science.

In *Optics,* the instruments are expensive and many of them can be only obtained by importation, and on that account we have deferred the purchase of them. A new Solar Microscope is almost indispensable. The old one procured at second-hand nearly forty years ago, is so inferior an instrument and with all so much injured, as to be scarcely decent for our use. The cost of a good instrument and appropriate set of slides is about one hundred and thirty dollars.

We also need a complete set of apparatus for demonstrating and illustrating the Laws and phenomena of *polarized light,* costing about one hundred and twenty dollars.

We ought also to have a Camera Lucida, lenses and prisms for the *Aberration,* the *Inflection* and the *Chromatism* of light; and Mirrors of different forms, for the illustration of the Laws of *Catoptrics.* These would probably cost not less than about one hundred dollars. A *Heliostate* is an instrument possessed by most of our respectable Colleges, and is essential in many optical investigations, but it is expensive, costing from 100 to 160 dollars, and not absolutely *required* though very convenient in *teaching of our pupils.*

In addition to these a *Magic Lantern,* with suitable appurtenances for exhibiting and explaining the phenomena of dissolving views, with slides for illustrating Astronomy and Natural History, & c. would greatly add to the interests of students in the department, but would be expensive and perhaps not indispensable, inasmuch as we have a small and inferior instrument, for ordinary exhibition. It would cost with slides from 150 to 200 dollars.

In Meteorology, we think it very important that we should be furnished with a complete set of the standard instruments that have lately been introduced under the direction of Prof. Guyot, at some of the principal points of observation through the United States.[3] These, I believe consist of Barometers, Thermometers, Hygrometers, Anemometers, and Rain Gauges, & c. which have been carefully constructed and compared with prime standards. I know not the cost but I presume not less than $100 or 150 dollars . . .

We thus make up an amount of *indispensable* 925.00

If we add other instruments named amounting to about $475 we have a total of $1400.00

In the above, I have not included magnetic apparatus, because a complete [set] almost involves the establishment of a magnetical Observatory, which would cost, as in the case of Williams College, about $1000. If the Chandler School should prosper, such an Observatory might at no distant day, be important for us. I do not deem it necessary at present.

If, however, we had the means, I should deem it of far greater importance that we should have a good set of Geodaetical instruments, to be used by our students, under the direction of the Professors, as well as by the Professors themselves, in excursions for Topographical surveys of our vicinity. These would be a good *Chronometer,* a *Portable transit instrument,* a good Theodolite, a portable Universal Instrument, a Zenith Telescope, such as is now used in the coast surveys for determining Latitude; and a portable Barometer and Hygrometer, suitable for making the necessary observations to determine the elevation of hills and mountains. Such a set would probably cost from $1000 to $1400, and would be a treasure to the College and its pupils.

---

3. Professor Arnold Guyot was a Swiss physical geographer whose Lowell Institute lectures in Boston in 1849 were translated into English (by C. C. Felton, subsequently President of Harvard College) to form the basis of a very popular book, Gould, Kendall and Lincoln, *Earth and Man* (Boston 1850). His publication in 1852 of what subsequently came to be known as the Smithsonian Meteorological Tables codified a vast network of weather observations in this country.

This added to the $1400, above, would make about $2800; still this amount, large as it is, and great as would be the benefit it would confer on our Alma Mater, when *added* to what has already been expended by us for Philosophical Apparatus, would fall considerably short of what already has been expended by Amherst for similar purposes.

Thus, my dear Sir, I have frankly stated to you, what I sincerely think to be very important wants in our department of apparatus, and will only add, that while I hardly dare flatter myself that we can immediately secure such a treasure, we shall be most grateful for the possession of any of the instruments which helped to make up whole. Sincerely yours, Ira Young.

Before the close of the special meeting of the Board of Trustees December 1852, all of the plans and requests outlined by Professor Young were voted and approved and the cobwebs of the past were brushed aside by the energies of this remarkable man.

Early in 1853 Professor Ira Young departed for Europe with his son Charles, who had drawn the diagram of the observatory. Charles Young at that time was an undergraduate and was later to become one of the most famous astronomers in America during his lifetime. Professor Young and son Charles visited London, Paris, Brussels, Halle, Munich, Geneva, and Edinburgh, working hard all along the way to discover what European universities were doing in the field of astronomy and natural philosophy and ordering the apparatus which had been so meticulously outlined to Dr. Shattuck. They left Europe in September of the same year and returned to oversee the completion of the observatory, which had been begun under the direction of Professor Hubbard in their absence. (Oliver Payson Hubbard was Professor of Chemistry, and it is interesting to observe that chemistry began to ride on the enthusiasm generated by Professor Young. With the success of Professor Young's memorials to the trustees, Professor Hubbard also started asking for equipment, and though his appropriations were numbered in the hundreds rather than the thousands of dollars, still a start was made in improving chemical instruction.)

As is obvious from the foregoing, Ira Young's strength lay

13. An early picture of the observatory. As actually built, it was all of brick, not with wings of wood as originally designed. The "old pine," a famous landmark of Dartmouth College, was fatally damaged by lightning in 1887, although it was not cut down until 1895.

in developing experimental departments of astronomy and physics. He was a man of good common sense, unaffected manners, and a pleasing personality, although he was apparently not a particularly inspiring teacher. A number of stories have been handed down about his inability to control the rowdyism of his students. For example, a member of the Class of 1835 reminisced:

> There was Professor Ira Young, white haired at 37 and dead at 57. His forte was mathematics and in that forte he was impregnable, but in the science of managing boys he was not so well up and they knew it better than he did. The darkening of the philosophical lecture room in order that he might reveal to the class a squirming world in a drop of water was the signal for such squirming on their part that he made haste to flood the room again with light, only to discover a most grave and composed assembly of interested youth. It is impossible to recall his amiable efforts to do us good without at the same time recalling such obstructions in his path.[4]

By early in 1858 it was obvious that Professor Young had some undiagnosed disease which sapped his vitality. He no longer dashed about the campus in his usual fashion, but was seen, as President Nathan Lord remarked, "Plodding his way to yonder dome with steps restrained and painful." Physicians decided that he needed to undergo an operation, a terrible ordeal in days before anesthesia, and on September 13, 1858, he died from the shock of this attempt to relieve his suffering.

---

4. The story about Professor Young's classroom discipline comes from remarks made by Charles E. Stevens to his 20th reunion and published in *The Dartmouth*, Vol. 6 (1884–85), p. 246.

Chapter 6

FOR THE GLORY OF GOD

Oᴺᴇ of the weaknesses of a small college with single-man departments is the complete absence of provision for continuity. Great teachers gather around them disciples who, although seldom as stellar as their masters, carry on traditions and, if nothing else, consolidate the original contributions and maintain some of the intellectual progress in an academic institution. But the size and character of Dartmouth a hundred years ago was such that not until a professor retired or died did the administration give active thought to a successor. Then, more often than not, the practical consideration of who could be persuaded to take his teaching load with a modicum of competency was more commonly a deciding factor than the identification of someone specially qualified to maintain and perhaps extend the achievements of his predecessor.

Ira Young was unique; any man asked to follow in his footsteps would have been hard-pressed even to maintain his achievements. But it seems almost as though President Lord went out of his way to find the antithesis to Young when in 1859 he appointed Henry Fairbanks as the Appleton Professor of Natural Philosophy. Young was the energetic carpenter's apprentice from Lebanon, New Hampshire, who earned himself his own education and went on to put Dartmouth at the forefront of physics and astronomy by literally designing, building, and stocking with instruments Reed Hall and the Shattuck Observatory. He raised the necessary funds himself, and like a magnet drew not only students but faculty to his lectures and star-gazing open houses. He was the first Dartmouth science professor of the modern era who believed in scientific knowledge for its own sake, unencumbered by its theological implications.

Fairbanks, in contrast, was the scion of a wealthy New England family, fired with the evangelistic fervor of fundamentalist New England Congregationalism, whose whole life was spent in trying to convert the wayward and whose commitment to natural philosophy was that of demonstrating "the glory of God." It is also probably true, though somewhat cynical to suggest, that President Lord, enmeshed in the desperate financial straits of the College, was eager to maintain a close relation with an heir to the Fairbanks wealth, particularly so since, as appears from the College records, Henry Fairbanks was willing to teach for little or no compensation.

In 1824 Erastus and Thaddeus Fairbanks had formed E. & T. Fairbanks & Company in St. Johnsbury, Vermont, to manufacture cast-iron stoves and ploughs. Thaddeus was the clever inventor and immensely successful designer of both, and when, in 1831, he invented the compound lever platform scale, his fame was secure for generations to come. While Erastus, the president of the organization, devoted some of his time to politics, Thaddeus continued as the epitome of the Yankee inventor, his fame spreading to such an extent that he was even knighted by Queen Victoria.

Henry Fairbanks was born on May 6, 1830, son of the inventive Thaddeus. Although probably apocryphal, the following story found in a newspaper at the time of his death illustrates the puritanical rigor of his Vermont upbringing: "When nine years old he rode seven miles and back each day to Lyndon Academy, and it always pleased him to recall the morning floggings; the master explained as he used the rod, 'If you haven't done anything to deserve it yet, you will before night.' Later, his father founded St. Johnsbury Academy, where Henry fitted for college."[1] He graduated from Dartmouth and was elected to Phi Beta Kappa in 1853.

---

1. The tale about Fairbanks as a nine-year old comes from a news-

Almost nothing is known of his undergraduate years, except what is contained in his own less-than-enthusiastic reminiscence of the daily chapel service:

> I remember the first morning of the spring term of that year [1850] going through the storm to that room [the chapel], the walls bare with the remains of partly erased charcoal sketches, the woodwork painted a dirty white, so sticky that it had been growing more dirty until in places it was black, all whittled and pencilled—President Lord and Professor Sanborn showing dimly through the smoke that made our eyes smart intolerably, and was the only sign of fire, for the snow that had sifted through the broken panes remained unmelted—students in cloaks that partly concealed their deficiencies in toilet having left their beds only six minutes before, rushing in at half past six; the Freshmen seated conveniently next to the door to invite the Sophomore rush; barbarian surroundings at a barbarous hour, doing the utmost to make barbarians, and this in the name and form of a religious service.[2]

As has been pointed out before, Dartmouth, like other colleges in New England, drew its faculty primarily from the clergy, and the natural route to either the ministry or a college professorship was through a theological seminary. Dartmouth was much ingrown at this stage of its development: drew its faculty almost exclusively not only from its own graduates, but also from among those who had obtained their postgraduate training from the Andover Theological Seminary. During Nathan Lord's presidency, three-quarters of the faculty were Dartmouth's own graduates, and two thirds of them, moreover, had proceeded from Dartmouth to Andover. It is therefore not surprising to find that young Henry Fairbanks on being graduated from the college went to Andover for this theological degree. He had interrupted his studies at Dartmouth in 1848 to go abroad and remained in Europe almost a year, and again while

---

paper clipping, the origin of which is not recorded, held by the Dartmouth College Archives.

    2. The description of the chapel service in the dusk of the morning is reproduced in L. B. Richardson's *History of Dartmouth College* (Dartmouth Publications, 1932, Stephan Daye Press, Brattleboro, Vermont), on p. 481 of Volume 2.

he was at Andover he traveled through Greece, Egypt, and Palestine; and it is said that on the way home he made an ascent of Mont Blanc, one of the first Americans to do so. He received his Master's Degree in Divinity in 1857 and returned to his native Vermont to become Superintendent of Itinerant Preaching for the Vermont Domestic Missionary Society.

It is obvious that Fairbanks' academic specialization did not bring him to the fore as a candidate to succeed Professor Young, whose death in 1858 left the teaching of natural philosophy and astronomy unattended. In fairness, however, it should be acknowledged that it was not uncommon for scientists at that time to be self-taught. The usual practice was to try out potential instructors of science and mathematics through apprenticeships as tutors before offering them professorships, especially if they had had no formal preparation. Hanover undoubtedly knew Henry Fairbanks as a person of an ingenious and inventive turn of mind, and certainly in later life, as he rose to be director and then vice-president of E. & T. Fairbanks & Company, he showed himself to be his father's son by holding many patents for inventions of interest to the company. Then, too, President Lord was not having an easy time either with his faculty or with his board of trustees. With Ira Young's death it appeared that no one of comparable talent was available to take his place, and also the chairs of astronomy and natural philosophy were separated, so there existed the added burden of recruiting two new professors, instead of only one.

Fairbanks was proposed by the Board of Trustees as Professor of Natural Philosophy, and President Lord agreed. Professor James Patterson, then Professor of Mathematics, was very eager to assume the duties of astronomy, and in a most unheard-of fashion he began to campaign for the position. He obtained a formal vote of the faculty in his support and even arranged for each of the undergraduate classes to submit a petition to the president favoring his selection. He also solicited sponsorship of his application from numerous benefactors of the College. President Lord, who wanted to appoint Professor Ira Young's son Charles, was forced to abandon the attempt and to yield to the mounting pressure in Patterson's behalf for the astronomy post.

14. Henry Fairbanks.

It is very apparent from the available records that life did not move smoothly around Professor Fairbanks. At times so much criticism arose concerning his lack of effectiveness as a teacher that it seemed as if action might have to be taken for his removal. His ability to maintain discipline was woefully lacking. Such tales as James Powell reports in a letter preserved in the Dartmouth Archives, dated September 24, 1864, were not uncommon:

> The recitation room is disorderly and it would defy Van Amburg with his menagerie of wild beasts to get up more disorder and confusion than is manifest in our recitation room before Prof Fairbanks. Yesterday afternoon the recitation room was a very bedlam. The boys filled their pockets with little stones, shot, crab apples crackers and whatever else was available and portable and carrying them into the recitation room, awaited it commencement when slap bang they would go and missiles would be flying about as thick as hailstones and fall upon the floor making a rattling like so much musketry.

Another story told about Professor Fairbanks was of the time when, in the middle of the night, the bell on Dartmouth Hall started ringing violently, and Fairbanks stormed up to the top of the building to catch the culprits at their prank. The bell was rung from a rope which extended down through two floors, but obviously the students had a lookout who sounded a warning for escape and by the time Fairbanks got to the top floor all he could see was the rope hanging out the window. He looked out and saw three students hanging on the end of the rope a few feet below the level of the sill, so he calmly took out his knife and cut off the rope. It is not recorded how far they actually fell, but apparently no one was seriously hurt—although the boys landed in a large burdock bush.

The records also show that Henry Fairbanks married Professor Noyes's daughter Annie while college was in session (on April 30, 1862). The students found out about it, and the morning of the wedding a great parade of them wandered all over the town and through the College buildings chanting, "Poor Annie, poor Annie, poor Annie."

Nor was all sweetness and light in Fairbanks' relations with the rest of the faculty. It is hard to imagine a modern board of trustees being asked to deal with interdepartmental squabbles about schedules and course times, but in the College archives there is preserved an interesting letter addressed to "the Honorable Board of Trustees of Dartmouth College" marked "Professor Fairbanks' Memorial, Professor of Natural Philosophy 1861" in which Fairbanks demands relief from his teaching load and complains about the division of labor between him and other professors.

In the autumn of 1862, Professor Patterson, perhaps in part stimulated by his successful political campaign for his astronomy chair, ran for Congress and was elected. With a desire not to lose the security of his position on the Dartmouth faculty, he petitioned the Trustees to let him hold his professorship during his two-year term in Washington and to instruct the faculty to provide the teaching he would be unable to give. The Trustees were obviously unhappy about this request but agreed to it with the proviso that his salary should cease while he was absent from the College. Not regarding this as any permanent kind of a solution, at their annual meeting in July 1863 the Board voted to abolish the Professorship of Astronomy and reunite it with the Appleton Professorship of Natural Philosophy under Professor Fairbanks, returning Professor Patterson to the Chair of Mathematics. In the storm which arose, Professor Fairbanks refused to accept the added labor entailed with the Department of Astronomy, and Professor Patterson, who had worked so hard to get the Chair of Astronomy, declined to accept the Professorship of Mathematics. Since it was much too late in the year to find any replacements, the Trustees had to back down and to ask both men to continue as before.

The battle was not at an end, however, because the Trustees voted at their August 1864 meeting that "here-after the acceptance of any civil office by any member of the Faculty, except the office of Justice of the Peace or any Town Office, shall operate *ipso facto* as a resignation of his position as a member of the Faculty."

The Trustees got themselves in trouble again by this vote

since in the fall of 1864 Mr. Patterson was returned to Congress
for a second term, and since they had in the meantime filled the
post of Professor of Mathematics, they could not use it as a pawn
as they had in their previous juggling. They came up with another scheme at their meeting in October 1865:

> *Whereas* the department of Natural Science in its various
> branches, is deservedly growing in public favor, and the more
> because of the rapid development of the natural resources of
> the country; and *whereas* it is the purpose of the Board that the
> College shall lack nothing essential to the broadest culture;
> therefore.—
>
> Resolved 1. That we hereby institute a new Professorship to
> be called the Professorship of *Natural History,* the incumbent
> whereof shall give instruction by lectures and otherwise in
> whatever pertains to such chair, aided by such cabinets and
> other apparatus as may be needful. And it is particularly re-
> ferred to the Faculty to consider whether the subject of *min-
> ing,* which is assuming such prominence in our country, and is
> becoming a matter of importance in our own vicinity, may
> not properly be embraced in the programme of this Professor-
> ship.
>
> And *whereas* the Board still deems it desirable, in accord-
> ance with the vote passed at their last meeting, to unite the
> Professorships of Astronomy and Meteorology with that of Nat-
> ural Philosophy, thus restoring the arrangement, which under
> former Professors has proved so convenient and successful; and
> *whereas* Professor Patterson who has officiated with marked
> ability in our board of instruction for eleven years past, by ac-
> cepting the office of Representative in Congress to which he
> was elected March last, has, according to a rule adopted by the
> Board at their meeting in August 1864, virtually resigned his
> place in the College Faculty; and *whereas* Professor Fairbanks,
> who for more than five years diligently and faithfully served
> the Board in the professorship of Natural Philosophy, has, on
> grounds which seem reasonable, especially on considerations of
> health, heretofore declined undertaking the duties of both
> chairs, therefore.
>
> Resolved 1. That the two professorships above named be and
> they hereby are united.
>
> Resolved 2. That Professor Fairbanks be transferred to the

new professorship of Natural History—the change to take place at the commencement of the next spring term.

Resolved 3. That we elect, at the present meeting, an incumbent of the united professorships, to be styled the "Appleton Professor of Natural Philosophy and Professor of Astronomy," his salary to be $1,300 per annum, and his services to commence at the beginning of the spring term.

The lack of an adequate salary associated with the new Professorship of Natural History was of no particular concern to Professor Fairbanks, and he accepted the position forthwith. The Board did go on to elect an Appleton Professor of Natural Philosophy and Professor of Astronomy, who is the subject of the next chapter, but before exploring that development it might be appropriate to chronicle briefly the remaining years of association between Henry Fairbanks and Dartmouth College.

Professor Fairbanks was probably willing to be shifted to Natural History because he really did not care very much what his responsibilities were so long as they did not entail more work. He was getting tired of the academic life anyway, and in 1868 at the age of 38 he resigned from the faculty, sold his house to Hiram Hitchcock (the donor of Hanover's Mary Hitchcock Memorial Hospital), and entered the family business in St. Johnsbury.

Two years later he was elected a Trustee of Dartmouth, a position he held until 1905. He was always considered to be the most conservative member of a conservative board, and he will be especially remembered in the annals of the College for his violent opposition to alumni representation on the board of trustees—a program which he blocked for many years, and finally hired lawyers to try to prevent the seating of such trustees after they were duly elected. His opposition, apparently, was completely on religious grounds as explained in part in a letter which he wrote to Professor John K. Lord at the time Lord was collecting material for his history of the College:

You recollect the declared hope of leaders in the Alumni movement of 1885 and later to shift the College from its Christian moorings. . . . Do you remember the article by Barrett, who

general'd the "New Dartmouth" movement, who wrote in the
Boston Advertiser of June 18, 1885, of the happy changes? He
says "The present Professors are a different type of men," "Some
of them are Episcopalian, some are practically Universalists.
Evolution is taught by the Prof. of Geology. . . ." "The moorings of the straight-laced orthodoxy are slipped, and the Theological learning of students is no longer the chief thought of
Professors." You remember those days and the effort to push
reconstruction on to changes so radical that they would have
alienated the best friends of the College. Moving more wisely,
the College has kept its friends, and has met all reasonable demands for progress, without sacrificing its own principles. It is
still Christian.[3]

Fairbanks was the last of the natural philosophers whose
commitment to the subject was through a zealous desire to ad-
vance conservative theology. God's handiwork on earth was to
be studied by man in order more fully to appreciate His omnip-
otence. To understand the intensity of Henry Fairbanks' commitment to this religious philosophy, one must read some of the
many tracts which he wrote as a leader in the Congregational
Church throughout his entire life.[4] His efforts were primarily
those of a missionary to those in New England who were not
adherents of the true religion, and he gave freely of both his
time and his fortune toward this end. He believed that "the

---

3. December 15, 1911; Archives of Dartmouth College.

4. In trying to appreciate the interest which Henry Fairbanks had in
the evangelistic missionary effort of the Congregational Church in New
England, considerable insight can be gained from three of his pamphlets.
One, *The Influence of Congregationalism upon Vermont,* was read before
the General Convention at Bennington and also before the Passumpsic
Congregational Club at Newport. It bears no date. A color-coded map
showing religious statistics in Vermont is contained in *Supplements to the
Minutes of the 91st Annual Meeting of the General Convention of the
Congregational Ministers and Churches of Vermont.* This session was held
at West Randolph on June, 1886, and its proceedings were published by
the *Vermont Watchman* and *State Journal Press* at Montpelier in 1887.
Much of Fairbanks' philosophy is contained in his address before the Boston Conference of the Evangelical Alliance, December 4, 1889. Entitled
The Needs of the Rural Districts, it was published by the Evangelical Alliance for the United States, 117 Bible House, New York.

strength of character born of independency, and developed by Congregational systems of education, is capable of fearful mischief when not controlled by principle or permeated by spiritual life."

Chapter 7

THE COMING OF RESEARCH

T HE sixth Professor of Natural Philosophy at Dartmouth College was the third generation of his family to hold that same position. Grandson of Ebenezer Adams and son of Ira Young, Charles Young was born in Hanover on December 15, 1834. Whereas his father had been prevented from embarking upon an intellectual career until he was 21, Charles was nurtured in the academic world from an early age.

It was soon evident that Charles was a precocious youth, and his father undertook the primary responsibility for his education. He entered Dartmouth at the age of 15 and was graduated in the Class of 1853, the same year that Henry Fairbanks received his degree. It is recorded that of the twelve terms of his college course, he spent three winter terms in teaching country school and, as already has been noted, the final term of his senior year on a trip to Europe with his father to purchase scientific equipment; so he was in actual attendance at the college only for two and two thirds years. For his first employment after college he taught Classics at Phillips Academy in Andover for two years. Throughout his life Charles was a contemplative and philosophic thinker. As an undergraduate, he was the first president of a missionary society called The Society of Inquiry, and in 1855 he entered the Andover Theological Seminary with the view of doing missionary work as his profession. However, after a year at the seminary he was offered an appointment as

Professor of Mathematics and Natural Philosophy at Western Reserve College in Hudson, Ohio. He accepted and entered upon his duties there in January of 1857. He had spent the previous fall in Hanover preparing for this assignment, and, one might surmise, wooing the charming Augusta Mixer of Concord, whom he married in August of that same year.

During the nine years that Professor C. A. Young was at Western Reserve, two important patterns of his life emerged. One was his interest in the science of astronomy, which surely must have been kindled by his early association with his father. The other was his commitment to research. His father had brought to Dartmouth a real awakening to the fascination of science, not only among the students but among the faculty and townspeople as well. His was the role of popularizing and making available to the nonspecialist an appreciation of science in general and astronomy in particular. Charles Young's role was to be different. Although he too was a popularizer of astronomy, his essential contribution really lay in research, in which realm his interest was so deep that eventually he left Dartmouth to go to Princeton, because the latter institution was more willing to allow him the time and facilities to discover new knowledge, at the expense of the teaching function of transmitting it.

In an examination of his activities while a professor at Western Reserve, one can see these patterns emerging. For part of the winter of 1858–59 and during the summers of 1864 and 1865, Young worked as an astronomical assistant to the United States Coast-Lake Survey on their problem of determination of longitudes. He also developed a time-signal service for the city of Cleveland. Although these were not major endeavors in themselves, they demonstrate that he was seeking experience in research and surroundings where apparatus was available to carry out investigations. As was true all over the country, his normal teaching pattern was disrupted by the Civil War. Having served on the campus at Hudson as an officer, he was commissioned a captain in June of 1862 and joined Company B, 85th Ohio Regiment. He was assigned to guarding prisoners of war until early October of that year, when he was mustered out and returned to his teaching duties.

It has already been noted that Professor Fairbanks was transferred by action of the trustees at their meeting in October 1865 to a new professorship of natural history. At that same meeting they elected Charles A. Young to Dartmouth as Appleton Professor of Natural Philosophy and Professor of Astronomy. It may be pointed out, in passing, that two thirds of the trustees who voted on this issue in 1865 were the same members who refused to support his appointment in November of 1859,[1] and yet there seems to be little evidence that his reputation either as a teacher or a natural philosopher had increased in the intervening six years. Be that as it may, with his move to Hanover his own scientific distinction and that of the College as well were almost overnight catapulted to the first rank, as Young turned his attention more and more to the field of spectroscopic and astronomical research and was able, through the availability of the ample income of the Appleton Fund, to equip both the observatory and his expeditions with the most modern of scientific equipment.

At the time Charles Young came back to Dartmouth, the most important astronomical problem was that of the nature of the sun, and Young was to become one of the world's greatest experts on this subject, about which he wrote:

> It is true that from the highest point of view the sun is only one of a multitude—a single star among millions—thousands of which, most likely, exceed him in brightness, magnitude, and power. He is only a private in the host of heaven.
>
> But he alone among the countless myriads, is near enough to effect terrestrial affairs in any sensible degree; and his influence upon them is such that it is hard to find the word to name it; it is more than mere control and dominance. . . . he is almost absolutely, in a material sense, the prime mover of the whole. To him we can trace directly nearly all the energy involved in all phenomena, mechanical, chemical, or vital. Cut off his rays for a single month, and the earth will die; all life upon its surface would cease.[2]

1. Opposition to C. A. Young from faculty and trustees is discussed on pp. 506 ff. in Richardson's *History of Dartmouth College.*

2. From C. A. Young, *The Sun,* 2nd ed. (New York, D. Appleton &

Young's particular contribution to the study of the sun was in the application of the then very new science of spectroscopy. This is the branch of physics which deals with the light emitted from luminous bodies, where the light is broken up into its constituent parts, and analyzed. The correlation of the wavelength and intensity of the light is a guide to understanding the nature of the body which emits the light. Young designed a particular kind of spectroscope which fitted on the end of a telescope, in place of the usual eyepiece, and was called a telespectroscope and he was invited to bring it along on an expedition sponsored by the *Nautical Almanac* office to observe a total eclipse of the sun in Burlington, Iowa, on August 7, 1869.

To understand Charles Augustus Young as a scientist one must of course read selections, at least, of his ten volumes of published works. But, a sense of the magic of the man's personality, of his ability to excite and to stimulate the people around him, and of his almost childlike enthusiasm for the world of science and astronomy in particular, one may turn to a letter he directed to the editors of the undergraduate paper, *The Dartmouth*. Filled with exuberance at having witnessed his first total eclipse of the sun, he wrote:

> From time to time it happens that when the moon is new she passes directly between the earth and sun, and then, if she is in the nearer portion of her orbit, her shadow strides across the land with unimaginable swiftness for more than two thousand miles an hour. Those who are in the track of this darkness behold a total eclipse—that grandest and most beautiful of all celestial phenomena. . . . A man seldom sees it more than once, and having seen it, always counts it an epoch of his life. To the astronomer, of course, such a phenomenon possesses, in addition to all the elements which excite the interest of others, the high-

Company, 1883), p. 11. He wrote the book, he says, "neither for scientific readers as such, nor, on the other hand, for the masses, but for that large class in the community who, without being themselves engaged in scientific pursuits, yet have sufficient education and intelligence to be interested in scientific subjects when presented in an untechnical manner; who desire, are perfectly competent, not only to know the results obtained, but to understand the principles and methods on which they depend, without caring to master all the details of the investigation."

est scientific importance. It is something not only to be admired
and wondered at, but to be carefully observed and studied; a
rare opportunity, which earnestly and wisely used, may ad-
vance him far in his investigations of the hitherto undiscovered
mysteries of the heavens . . . About ten minutes before total
obscuration, the sky appeared much as during a heavy thunder-
storm, and the strangeness of form in all small shadows, the
peculiar pallor of companion's faces, with the evident uneasi-
ness of not a few of the spectators, combined to give a very
peculiar impression not easily to be forgotten were it not for
the stranger sensations yet to come.

Little by little the silver thread narrowed, till all at once
there was a flash of darkness; night was upon us, and the stars
leaped into sight. The moon hung in the sky, dark, though not
absolutely black. Fastened upon her edge there seemed to be
several stars, of color like the planet Mars; around her was the
mysterious corona, formed of rays extending in all directions,
of varying lengths, like slender streamers of the Aurora Bore-
alis . . .

I cannot tell what it was that made the spectacle so impres-
sive; the attempt to put it into words evaporates it. I only
know that its effect was overwhelming; and even now were it
not for recorded notes, the memory of the few preceding minutes would be lost entirely, completely blotted out by the un-
speakable, uncommunicable glory I had seen, as if I had been
admitted to the immediate presence of the Creator. It was a
moment worth hours, perhaps years of ordinary life. Everything
was still as death except the click of the chronograph, and the
occasional snap of the photographic apparatus; but when the
sun burst out again, shouts filled the air, not from the astronomers of course, but from the crowds in the streets around.[3]

A picture of the *Nautical Almanac* party stationed at Burlington, Iowa, is shown as Figure 15.

Scientifically, Young made two important advances as a result of his observation of this total eclipse. Previous to going on
the expedition, he had tested his spectroscopic equipment by
observing the spectrum from the chromosphere, the outer lay-

---

3. *The Dartmouth*, Vol. 3, No. 8 (September, 1869).

15. A photograph of the astronomers at the Burlington, Iowa, eclipse on August 7, 1869. The telescope to the far left of the photograph looks identical to the English telescope which the Wheelock brothers arranged to acquire in England and which is illustrated as Figure 5, above. The second woman on the left is Maria Mitchell of Vassar College, visiting the site to have her picture taken with one of her fairest pupils. The large telescope in the middle was that used for photographic purposes under the supervision of Professor Mayer of Lehigh University. Professor Young and his assistant, Charles Emerson, who succeeded him as professor, are situated at the right end of the hut, and the spectroscope can be seen attached to their telescope. The individual out in the open may well be Professor J. H. C. Coffin of the *Nautical Almanac* office, whose duty was to tend primarily to the determination of time and, also, to make, during the totality, observations with a moderate-power telescope on the prominences and the corona. To the far right is Professor B. A. Gould of Harvard, who was searching for intramercurial planets with a five-inch telescope of low-magnifying power, and in the foreground is the all-important chronograph magnet which clicked off its exciting seconds.

16. The tele-spectroscope designed and used by Professor C. A. Young at the Shattuck Observatory, attached to the 9.4-inch telescope.

ers of the atmosphere of the sun, and had discovered a particular line not previously reported, which he called the 1474 line
of the Kirchhoff scale (which in modern terminology would
have a wavelength of 5,317 angstroms). He identified this line
as probably due to iron. During the total eclipse he was able to
demonstrate that this particular line came truly from the
chromosphere and was therefore characteristic of the sun and
not of the earth's atmosphere.

One of the problems during the eclipse was to determine
precisely the time of contact between the moon and sun images.
Just before the eclipse he thought of setting the spectroscope
radially with the sun's disk and of watching the bright sodium
D-lines shorten as the moon approached. Before contact the
lines would, of course, be of constant length, but as contact approached, the lines would begin to shorten and disappear on
contact. The method was accurate for determination of this
time. He found that during the eclipse his observation agreed
with other methods being used by other members of the expedition. Young suggested that this technique could be applied to
the determination of the contact time in the transit of planets,
particularly having in mind that there was to be a transit of
Venus in 1874.

Back in Hanover by the end of August, Professor Young
set himself to the task of writing up in detail for the scientific
community his observations during the eclipse, including a description of his spectroscope, which had proved to be such an
outstanding success. He redesigned the instrument and with it
continued to make observations of the sun's chromosphere, and
what he called protuberances (now called prominences). He was
very interested in the newly developing techniques of photography, and gained international note by succeeding on September 28, 1870, in obtaining successful photographs of solar prominences. This was really a remarkable feat and a tribute to his
skill, considering the fact that photographic technology had not
advanced beyond the slow, wet-plate process; it required an
exposure of three and a half minutes to obtain the details which
were discussed in a spectroscopic note issued as a special supple-
ment to the *Journal of the Franklin Institute*. Of course, photo-

graphic reproduction was not yet possible in publications of this sort, and what were published were not the actual photographs but drawings of them. With the proper modesty of the time, he wrote, "As a picture, the little thing amounts to nothing, because the unsteadiness of the air and the maladjustment of the polar axis of the equatorial caused the image to shift its place slightly during the long exposure of three and a half minutes which was required, thus destroying all the details. Still, the double headed form of the prominence is evident, and the possibility of taking such photographs is established."

As is obvious from the quoted section of Professor Young's text in *The Dartmouth,* his whole research plan after the Iowa expedition was to get ready for a similar series of observations in Spain in December of 1870. Between these two expeditions, Young and his assistants were very busy improving their techniques and designing a remarkably efficient new spectrometer of an optical quality so excellent that two generations of physicists were to use it at Dartmouth with unexcelled results. Basically in achieving this improved instrument, he redesigned the optical system so that the light was sent through his prisms twice, and the apparatus was made extremely light, in order that it could be easily transported. The eyepiece was arranged for optimum convenience of visual observation. He published several papers describing the details of the new spectrometer, so that others could be constructed, and he passed the designs over to the optical manufacturer, Alvan Clark & Sons, for marketing. In one account, after describing with great enthusiasm the operation of his instrument, he does somewhat wistfully remark that "the protuberances are so well seen . . . that it is even possible to photograph them, though perhaps not satisfactorily with so small a telescope as the one at my command."

The 1870 total eclipse occurred on December 22, almost at noon. And the expedition to which Professor Young was attached had gone to Jerez to make its observations. As was the case with the first expedition, Professor Young wrote to *The Dartmouth* about his experiences, providing a travelogue of his trip there, as well as tracing the features and events of the eclipse. In his report of the eclipse he wrote

Just at the commencement of the totality I made an observation, which was wonderfully beautiful to see, and which I
think has important theoretical bearing. The slit of my spectroscope was placed tangential to the sun's limb, just at the
base of the Chromosphere . . . suddenly, as the last rays of the
solar photosphere was stopped out by the moon, the whole field
of view was filled with countless bright lines—every single dark
line of the ordinary spectrum, so far as I could judge in a moment, was reversed, and continued so for perhaps a second and
a half, when they faded out, leaving only those I had at first
been watching.[4]

The sudden increase in intensity of the spectrum was described by Young in several scientific publications. This outstanding discovery, with which his name is still associated, is
now called the "flash" spectrum. It had been predicted on theoretical grounds but never before had a spectroscope been of
sufficient excellence to observe it. Not until 1896, when the first
photographs of the flash spectrum were obtained, was the controversy generated by his claim to have observed it as a simple
reversal of the Fraunhofer spectrum settled to Young's credit.

Professor Young had been preparing for a transit of Venus
observable in China for many years. His method (invented during his first eclipse observation in Iowa) of measuring the contact of the moon with the sun's disk by observing the length of
the spectral lines in his spectrometer had immediately suggested
to him an accurate method of determining contact between
Venus and the sun. It may seem strange, therefore, that for all
the preparations he made for the expedition to Peking to observe
the transit of Venus on December 8, 1874, Professor Young is
remarkably silent about the event. There are no articles by him
in *The Dartmouth* about it, nor any scientific papers of his describing the results. The reason is simple. The observations
which the expedition made clearly showed that measuring a

---

4. The articles written by Professor Young for *The Dartmouth* are
contained in Vol. 5, No. 3 (March, 1871), pp. 87–90, and called "A Bit of
Foreign Correspondence," dated "At Sea, December 1, 1870; Gibralter,
December 2, 1870; and December 7." His description of the eclipse itself,
also in Vol. 5, is called "A Letter from Spain," contained in pp. 57–62.

transit of a planet across the sun's disk was not a reliable method for determining the distance between the earth and the sun.

It is too bad that Young did not write up this expedition in narrative form in the way he wrote up the other ones, because apparently he had a very exciting time.[5] Three days after the successful observation of the transit and before the reduction of the data showed that this method was not as successful as had been hoped, the Emperor of China fell ill with smallpox, and the American legation with great haste and secrecy whisked the members of the expedition away, before the superstitious populace could credit them with sorcery in having somehow transferred the spots they were dealing with on the sun to the Emperor's face! Their seventy-mile journey to the coast, which was made in carts traveling mostly at night, was apparently as exciting as it was dangerous. They did, however, manage to flee from China with their photographs and their telescopes safe and sound.

It is characteristic of a scientist that as a creative man he is always searching for something better and, hopefully, more productive. One finds throughout Young's writings while at Dartmouth constant references to the inadequacies of his equipment, implying a desire for bigger and better instruments. He obviously pushed the College relentlessly to provide better telescopes, and in 1872 a private subscription of almost $5,000 was raised for the purchase of a new 9.4 inch instrument. This telescope is still in use in Dartmouth's Shattuck Observatory, and the tailpiece end, with C. A. Young's telespectrometer attached, is shown in Figure 16.

Young was a vigorous and dynamic research worker, and despite all the affection he felt for Dartmouth, his real goals in life were focused on research astronomy. Thus when in 1877 Princeton offered him a research position without any required teaching duties, and with the promise of a new 23-inch refractor, plus all the money he needed to create an even more powerful

---

5. Reference to the events surrounding the expedition to Peking are given in a paper written about C. A. Young by Professor John M. Poor, in *Popular Astronomy*, Vol. 16 (1908), p. 281.

solar spectroscope, he could not resist the temptation. He left
Hanover for Princeton, where he stayed until he was 70 years
old.

Professor Young's departure from Dartmouth coincided
with the advent of a new era in the behavior of professors of
science, something that was, indeed, to prove characteristic of
the educational scene all over the country. Professors no longer
would be willing to spend their lives exclusively in the classroom; a growing competition developed between teaching and
research for the professor's time and energy. Young's situation
was being duplicated all over the United States. These were the
first signs of the modern awareness that teaching must go hand
in hand with the acquisition of new knowledge. Dartmouth was
experiencing serious financial problems; it could not match the
offer from Princeton even if it wanted to. Yet Young had put
the college in the intellectual limelight, not only of this country,
but of the world. Students and faculty alike were caught up in
the excitement of new knowledge, and it was obvious to all that
an education at Dartmouth was now much more than training
for the ministry or for teaching. Young stated very precisely
what the college must do if he were to stay. First, he asked to be
relieved of his duties as Professor of Natural Philosophy, so that
he could spend full time as Professor of Astronomy. Second, to
avoid the constant annoyance of competing for salary and salary
increases, he asked that the professorships of Astronomy be supported by an endowment fund sufficient to relieve the college
of using any other funds for this purpose. And, finally, he asked
that a fund similar to the Appleton Fund for Natural Philosophy be set up specifically for apparatus in astronomy. The
Board of Trustees agreed in principle to these requests, but the
college was so deeply in debt that their fulfillment was hope-
less.[6]

---

6. In an attempt to keep Young at Dartmouth, the Board of Trustees
agreed to raise a subscription of $100,000 to endow a chair in astronomy,
and one of the stipulations President Samuel Bartlett made in coming to
the college was that this fund be created by the trustees. After Bartlett had
taken office, he discovered that there was a debt amounting to $125,000
about which nobody had informed him. He soon discovered, also, that not

From what has just been said, it would be very wrong to jump to the conclusion that Professor Young was not a great teacher. He was in fact a beloved and effective teacher of undergraduates. After he left Dartmouth he developed the nickname "Twinkle," and this so fitted both his friendly nature and his warm personality that its use was immediately adopted by many of those at Dartmouth who had known him during his years on the college's faculty or who were to come to know him later, when he retired to Hanover to spend the last years of his life in his old home.[7] One might also get the idea that he spent most of his time being a professor of astronomy and neglected his natural philosophy. This was not at all true, for he not only produced a good many instruments of novel design, which one could consider to be either astronomical or physical, but also found pleasure in inventing equipment for the physics laboratory and demonstration. He built and published a description of an automatic mercurial pump, a gravity escapement for a

---

only did he have to raise that amount, but that the trustees, although they had agreed to raise the $100,000 for the chair in astronomy, were taking no action in that regard. Bartlett never forgave the trustees for this, and in his last annual report, years later, he reminded them of that fact, and commented that he would never have accepted the presidency of the college if he had realized how bad in fact the situation was.

7. An indication of the effect which Young's proposed removal to Princeton had on the students is provided by a notice in the Boston *Journal* for February 14, 1877:

The following report, relative to the proposition made to Professor Young to accept the chair of astronomy at Princeton College, which expresses the feelings of the students, has been presented to the Trustees of Dartmouth College.

"Whereas, one of our faculty, Professor C. A. Young, is considering the offer made to him to occupy the chair of astronomy at Princeton, and

"Whereas we the students of this College, conscious of the great and irreparable loss which his acceptance would bring us, and,

"Whereas we gratefully remember his untiring zeal in behalf of the College and all connected with it. as well as his worldwide fame, therefore

"*Resolved,* that while recognizing also her financial embarrassment, we respectfully pray that if possible Professor Young be retained in his present position."

pendulum clock, and wrote a "Note on Recurrent Vision,"
which he describes in part as follows:

> In the course of some experiments with a new double plate
> Holtz machine belonging to the college, I have come upon a
> very curious phenomenon which I do not remember ever to
> have seen noticed. The machine gives easily intense Leyden
> jar sparks from seven to nine inches in length, and of most
> dazzling brilliance. When in a darkened room the eye is
> screened from the direct light of the spark, the illumination
> produced is sufficient to render everything in the apartment
> perfectly visible; and what is remarkable—every conspicuous
> object is seen *twice* at least, with an interval of a trifle less than
> one quarter of a second—the first time vividly, the second time
> faintly; often it is seen a third and sometimes, but only with
> great difficulty, even a fourth time. The appearance is precisely
> as if the object had been suddenly illuminated at first bright,
> but rapidly fading to extinction, and as if, while the illumina-
> tion lasted, the observer is winking as fast as possible.[8]

If one defines as physicists those educated and trained in
that branch of the physical sciences which deals with the laws of
force, matter, and energy, and the natural philosophers as those
educated in the classics, theology, and philosophy, who turn
their attention to the nature of physical laws, then Charles
Augustus Young was the last and greatest of Dartmouth's natural
philosophers. Not only did he make tremendous contributions
to our understanding of the nature of the universe, but he
thought deeply and wrote eloquently on the interaction of science and religion. He authored many volumes describing the
details of physics and astronomy in simple terms understandable to the common man. He was a popular and scholarly pro-
ponent of the school of thought, espoused among many sectors
of present-day society, which believes that science should be an
integral part of our culture and that the mark of a truly educated man is to be as conversant with science as with the hu-
manities. He was a true colossus, bridging the old with the new.

---

8. *American Journal of Science and Art,* Vol. 3 (March 1872), p. 1.

17. Professor Charles A. Young is best remembered puttering with his telescopes, which he could never bear to have idle. This picture, taken toward the end of his life, shows him in a typical situation, with one of his solar spectroscopes attached to a small telescope.

Mention has been made of a number of times when Young
went off on expeditions for months at a time. Several have not
been touched upon. After leaving Dartmouth, his life was as full
of journeys as before. He also spent time counseling the government on educational problems of the United States Military
Academy and on scientific problems for the Coast and Geodetic
Survey, as well as on special matters strictly in astronomical
areas. Because of his popular writings, his reputation as a scientist was as great in the world at large as it was among his fellow
scientists. Poems were written about him and to him. A 13,187-
foot peak close to Mount Whitney in California was named in
his honor,[9] and his fame had not been forgotten as late as the
Second World War, when a Liberty ship launched in Portland,
Maine, was named *The Charles A. Young* in his memory.

---

9. An interesting letter in the Dartmouth College Archives calls attention to Mount Young. This letter, written in 1934 from a J. H. Czock,
reads:

> While doing some climbing in the vicinity of Mt. Whitney this
> summer, we made a trip to the summit of Mt. Young, and found the
> original record of the naming of the mountain. Noting Professor
> Young's former connection with Dartmouth. I thought you might find
> a passing interest in the enclosed copy of that record. It is written in
> pencil on ordinary note paper and was enclosed in a tin can. The
> monument referred to is a carefully built cairn so constructed as to
> give the record can the maximum protection, which accounts for its
> excellent state of preservation.

> To make it more interesting, the second entry on the record on the
> reverse side was made by a party of three men on July 17, 1902.
> The third entry, our own, Mrs. Czock and myself, is dated July 24,
> 1934.

> Mt. Young can be found on the Mt. Whitney, Calif. quadrangle of
> the U.S. Geological Survey. The record and "monument" were found
> on the peak elevation, 13,187, marked "Mt. Young." and more nearly
> two miles west of Mt. Whitney. The peak N.W. of Whitney is not
> named (elevation 13,493) and careful search of its summit failed to
> disclose any record or remains of a cairn.

The paper record of naming the mountain reads, "Know all men! that I
hereby on the 7th day of September, 1881, do name this mountain '*Young,*'
in honor of Prof. *Charles Young*, now of Princeton and formerly Prof. at
Dartmouth College—in witness whereof I have hereon erected this monu-
ment as a perpetual memorial. Situation N.W. of Mt. Whitney, distance
about three miles, about N. of Mt. Hitchcock and about two miles distant,
Fred H. Wales."

Professor Young always considered Hanover as his home, a place of retreat and for vacations. It was therefore natural for him to return there on his retirement from Princeton. Although his health was not good, he spent the remaining three years of his life contentedly surrounded by his family and many friends. By one of those strange quirks of fate, on January 3, 1908, as a total eclipse of the sun made contact near the Caroline Islands in the South Pacific and tracked itself out across the ocean to end in Ecuador, Charles Augustus Young was "admitted to the immediate presence of the Creator."[10]

10. The quotation is from Young's own description of his first total eclipse of the sun, above, p. 72.

The Astronomer[11]
Robert Bridges

The destined course of whirling whirls to trace:
    To plot the highways of the universe,
    And hear the morning stars their songs rehearse,
And find the wandering comet in its place:
This was the triumph written in his face
    And in the gleaming eye that read the sun
    like open book, and from the spectrum won
The secrets of immeasurable space!

But finer was his mission to impart
    The joy of learning, the belief that law
    Is but the shadow of the power he saw
Alike in planet and in throbbing heart
    The hope that life breaks through material bars,
    The faith in something that outlives the stars!

---

11. From the New York *Sun,* December 9, 1904. In replying to Professor Young's thanks, Bridges wrote, "It is most kind of you to have written me about the sonnet. I tried to express in it what so many of your hundreds of pupils feel, and it is gratifying to know that you have not forgotten us individually."

Chapter 8

"A GENERAL UTILITY MAN"

"CHARLES FRANKLIN EMERSON was born in Chelmsford, Mass., September 28, 1843; his father was a farmer in comfortable circumstances, the son was taught that work is honorable and that there was little danger of having too much of it, if rightfully regulated and interspaced with pleasant recreation; in this atmosphere he learned to love work, and considerable responsibility in management and control was early entrusted to him; beside attending the public schools he was sent to Westford Academy under the Principalship of John D. Long afterwards Governor of Massachusetts and later Secretary of the United States Navy; in 1862 he taught his first school, an ungraded one, and continued teaching every winter and sometimes into the fall, till he graduated from College. He completed his preparation for college in 1863 at New Ipswich Appleton Academy under Mr. E. T. Quimby called to the Professorship of Mathematics at Dartmouth the following year.

"In response to letters from Prof. Quimby and Pres. Smith, he came to Hanover . . . Feb. 17, 1865, to enter Dartmouth, and has since made Hanover his home, having been continuously connected with the College."[1]

The author of the above is Professor Emerson himself, in

---

1. Manuscript copy in the Dartmouth College Archives.

an address commemorating the fiftieth anniversary of the Dartmouth Scientific Association, given on February 16, 1920.

As has been seen in the previous chapter, Charles A. Young was a spectacular genius whose accomplishments were such as to make the position of anyone destined to follow him difficult at least. Emerson was the man whose lot it was to be judged against his brilliant predecessor, and at the conclusion of the autobiographical sketch here cited, he himself observes: "From the above enumeration of details one is compelled to infer that the work of this member of the Founders has been of the nature of a general utility man in the College and Town, with very little time left for specialization in any branch of science." Yet despite this frank confession on his part of a lack of professional distinction, Emerson was good for the college, he was good for the science of the College, and he was a good man to personify the transition into the modern age of his discipline: the Appleton Professor of Natural Philosophy who was "translated" to Appleton Professor of Physics. He was good in many things without excelling in any. As an undergraduate he had demonstrated his competence in the classics by being invited by the faculty to deliver the Greek Oration in his junior year, and he delivered the Latin Salutatory when he was graduated. He was also called upon for an English Oration in his senior year. His athletic prowess was such that as a senior he taught gymnastics in the just-built Bissell Gymnasium, and with such success that the College sent him to summer school at Yale to learn more about physical education, from which program he returned in the fall to be instructor of gymnastics at Dartmouth.

Following graduation from Dartmouth with the class of 1868, besides being appointed to his instructorship in gymnastics, he was elected a tutor in mathematics, a subject he taught to freshmen thereafter for a period of ten years. Besides this, he was an instructor of mathematics and of natural philosophy in the New Hampshire College of Agriculture and the Mechanic Arts, until that institution moved from Hanover to Durham in 1874.

If one considers the tremendous activity of C. A. Young at

the time Emerson was just beginning his career, especially his concentration on research in astronomy and his frequent expeditions of one sort or another, clearly as far as Dartmouth was concerned somebody had to be on hand to take over the routine teaching, particularly in natural philosophy, and generally to keep things running when Professor Young was away from Hanover. It was easy for the College to turn to their young tutor in mathematics to provide this back-up role. Furthermore, one gets a feeling that for Young, who was an ardent believer in sports himself, the combination of a mathematician, natural philosopher, and instructor in gymnastics was most appealing as an assistant.

Charles Emerson accompanied Young in 1869 on his first eclipse expedition to Burlington, Iowa, and later in 1871 to Sherman, Wyoming, the highest point on the Union Pacific Railroad in the Rocky Mountains, to make spectroscopic observations. They studied the condition of the solar corona, and subsequently, by comparing their sunspot records with magnetic observations made simultaneously at Greenwich, they demonstrated a connection between solar storms and variations of terrestrial magnetism. In the following year Emerson was appointed Associate Professor of Natural Philosophy and Mathematics. He basked in Young's greatness as long as the latter was at Dartmouth.

The year after Professor Young left Dartmouth, Emerson was made Appleton Professor of Natural Philosophy. With the passage of time during his tenure, the new physics started overwhelming the old, new observations crowded in to cast doubt on the completeness of the classical concepts of Newton, Boyle, and Faraday, and the moderns of the day resented the philosophic implications of the discipline's designation. Institutions in this period therefore widely adopted for their natural philosophers the new title of physicists, and Dartmouth, following suit, in 1893 changed Emerson's position to Appleton Professor of Physics, the post he held for six years until he was made the first dean of the college.

The very day Young resigned from Dartmouth a new president accepted the leadership of the College. It was therefore

under President Samuel Bartlett that Professor Emerson took
over full responsibility for the Department of Natural Philosophy. Two men could not have been more different. Emerson
was a slow, kindly person with an active interest in men and
their affairs, while President Bartlett, almost from the day of his
inaugural address, plunged the College into the most acrimoni-
ous controversies.

> Dr. Bartlett possessed traits which were less [than] attractive
> and which made his task harder and his successes less impres-
> sive than they ought to have been. His demeanor was a liabil-
> ity; his manner was sharp, brusque, and even overbearing, and
> served to antagonize those whom it was highly desirable that
> he should conciliate. So clearly did he see the validity of the
> conclusions which he derived from the application of logical
> processes to problems under examination, that he could never
> understand why slower and less logical minds failed to accept
> those conclusions as quickly as or as completely as he did him-
> self. In a man of his keen mental powers, intolerance of stu-
> pidity was perhaps excusable, but he was slow to discover that
> such intolerance might, upon some occasions, advantageously
> be concealed. . . . These qualities made the course of Dr. Bart-
> lett more difficult than it should have been. Perhaps they may
> be summed up by the statement that he had little acquaint-
> ance with the art of managing men.[2]

With such a man at the helm, it is entirely understandable
why Professor Emerson, having a deep devotion to the College
and possessing, too, the basic personality of a peacemaker, should
have found himself more and more immersed in the personal
relationships between faculty, administration, and students, and
less and less with the highly competitive and fast-moving fields
of research physics and astronomy.

By 1881 the situation with President Bartlett had become
so critical that the Trustees set up a full-fledged trial, which
convened in the summer of that year. The Trustees demanded
that the President appear, to show cause why he should not be
dismissed. Professor Emerson was among the seven members of

---

2. Richardson, *History of Dartmouth College,* p. 590.

the faculty who openly supported the President, believing that he was in fact an able administrator and that it was primarily a quick temper and an intolerance of less able men that had brought on the contentions and controversies of his administration.

President Bartlett defended himself with such skill that the official charges against him were dissolved, but an uneasy truce descended upon the whole college community. Tension never really disappeared while Bartlett was President, to the point that Emerson was prompted, as late as 1890, to cry out, "Why can't we live in peace? Dartmouth has all it can do to thrive without being forever in a fuss. This wrangling is tiresome and will wear us all out and ruin the College. Give, oh give us peace."[3]

During this period Emerson, to quote again from his autobiographical sketch, "was called upon to help out in the Chandler Scientific School and in the Agricultural College, especially in physics, as those departments had no apparatus. There were months and even years when he was confined in the classroom between 20 and 30 hours a week giving instruction in physics and astronomy. Moreover much of the time he was on various committees of the faculty . . . and at one time he held fourteen offices in the College, town, and church." As his good friend Professor John K. Lord said of him after he died:

> He was a prominent and active citizen interested in everything that looked to the welfare of the community . . . his industry was pre-eminent; he never shirked work and as he did not acquire rapidly he knew that the cost of acquisition was work. He was fond of details, unwearied in their arrangement, and sometimes seemed to enjoy them more than the whole to which they belonged. His desk was always a model of neatness and his painstaking attention to least matters and his desire to secure all possible facts produced the almost perfect edition of the general catalog of the college.
>
> His sense of duty equalled his industry. He did not pursue careful speculation as to conduct or the issues of conduct and

3. Ibid., p. 612.

he did not make excuses when he felt something ought to be
done. When others seemed to draw back from hard or disagree-
able tasks which some one had to do, he came forward and took
up the labors which might justly have been shared. He was one
of those men who keeps a community going by the individual
assumption of common burdens. Having assumed an obliga-
tion he loyally carried it through and in the life of the community his sense of duty took the place of balanced argument
that might offer an excuse for withdrawal.[4]

One of the difficulties in giving Emerson credit for his contribution in natural philosophy was his self-effacing nature.
Not that this was an attitude he consciously assumed, but he
was so immersed in helping his younger associates and the society in which he moved that he left few records of his own activities. Professor Emerson, moreover, was a very modest man,
and although, as will be seen, he made a number of excellent
contributions to science, they rarely appeared under his name.
Usually he allowed them to be published by his younger assistants, who carried out the studies under his direction. All we
have are bits and pieces. For example, there exists a broadside
announcing "Martha's Vineyard Summer Institute. The Third
Annual Session of five weeks will begin Tuesday July 6, 1880 at
Cottage City, Mass., (formerly Vineyard Grove, Oak Bluffs, etc.).
Department of Astronomy will be under the charge and personal
instruction of Charles F. Emerson, A.M., Professor of Natural
Philosophy and Astronomy and Director of Shattuck Observatory, Dartmouth College, Hanover, N.H."[5] It is recorded, furthermore, that "the course marked out in this department would
be more appropriately styled 'Physics of Astronomy,' for it is
proposed to treat more particularly of those subjects in which
there is a most intimate connection between Physics and Astronomy; the most interesting field in modern Astronomy where
physical principles have such important applications, not only
in instruments, but also in the deduction of observations resulting from their use. Emphasis will be given to the proper under-

---

4. *Dartmouth Alumni Magazine*, Vol. 15 (1922–23), p. 218.
5. Dartmouth College Archives.

standing and appreciation of Optics, Thermotics, Electricity, and Chemistry, as applied in the Telescope, Spectroscope, Photometer, etc., and in Photography. The instruction will be given in the form of familiar lectures, accompanied by a short quiz on the subject matter of the previous lecture . . .

"For the use of the members of the department there will be two small portable telescopes, a large spectroscope with twelve prisms, and two small ones; many of the lectures will be illustrated by projections on the screen from the Porte-lumiere and Stereopticon, if the number of students warrants the expense."

There is also indication that Professor Emerson's reputation as a scientist extended considerably beyond the Dartmouth campus. A Boston newspaper of September 17, 1881, printed the following:

> The country presented a most peculiar appearance all day yesterday, there being no time from morning until noon when blue sky was to be seen or greenish, yellowish color did not prevail.
> At Fitchburg, it was so dark that the schools were dismissed, the smoke so thick it could be smelt. At Fall River schools were dismissed and in all the mills gas was lighted. The peculiar light that was intensified was red and green.
> A religious group at Worcester congregated at a schoolhouse to await what they considered the final end of the world. At Millford, N.H., the frogs croaked and crickets chirped as in early evening, many persons complained of a dizzy feeling. C. F. Emerson, professor of physiology (sic) and astronomy at Dartmouth College said that the phenomenon was caused by something in the atmosphere which absorbed the short and longer wave lengths, leaving only those which gave the colors yellow and green. He thought it might be owing to the smudges of fir and pine trees together with the smoke and forest fires in Canada.[6]

On two occasions Professor and Mrs. Emerson went on trips to Europe, and their stay in Leipzig in 1883–84 results in our knowledge in considerable detail of the physics courses

---

6. Quoted by the Lewiston Maine *Sun*, June 7, 1941.

which Professor Emerson was teaching in Reed Hall at that
time. The standard textbook in physics in the late 1800s used
throughout the United States was an English translation of a
book written by Ganot entitled, in its English translation, *Elementary Treatise on Physics, Experimental and Applied*. This
book went through a great many editions and set a pattern for
physics courses for at least two generations. While Professor
Emerson was in Leipzig he had time to find out what German
professors in that university were teaching, and he went carefully through his copy of Ganot, interleaving changes, corrections, interesting additions, and the latest measurements, in
many cases adding several pages of text. His wife, who was artistically talented, provided added illustrations of various pieces
of equipment not to be found in the textbook.

This single volume of Ganot's text with all the interpolated
material is rebound into two volumes. Not only does the work
contain a great quantity of added material, but also numbers,
corresponding to the number of the lesson where he stopped
lecturing, questions to ask the class, sections to omit, and various numerical answers worked out. One of the results is to show
clearly how slavishly he followed the book, and how, once having had an added interesting detail, it probably was repeated
year after year.

> In the late 70's and early 80's [Emerson] gave many illustrated
> lectures at teachers' institutes and general lyceums in N.H., Vt.,
> and Mass. using the extensive apparatus of the Appleton Fund,
> showing the development of science, especially along electrical
> lines, exhibiting the telephone, the electric light and the
> electric motor, thus he called the attention of many young people to the desirability of a College Education.[7]

The manuscript of one of these lectures is owned by the
Dartmouth College Archives and shows us clearly and rather
dramatically the state of technology and engineering in the
electrical field in Emerson's day.

Most interesting to us today is not so much the state of the

---

7. From Emerson's fiftieth anniversary lecture to the Dartmouth Scientific Association.

technology at the particular time when Emerson was talking, but, rather, his predictions of what might happen in the near future. Thus at one point he says, "the telephone, in its practical form, is scarcely five years old, and yet even now it is almost indispensable to the energetic businessman in our large towns and cities. I feel very confident that within the next five years we should be able to carry on conversations with friends in Mass. without leaving our homes."

The electric generator was just beginning to be introduced as a commercial machine and Emerson, in a lecture, speculated about its future:

> The transmission of power in the form of Electrical Energy to a distance of a hundred miles or more through copper cables without great loss, has been satisfactorily demonstrated, and the prediction is made that within the next decade some of our factories will be operated during the day and our houses lighted at night by Electrical Energy . . .
>
> The general distribution of Electricity will be needed not only for Electric lighting, but for a hundred other purposes of which we now hardly dream; when it is possible to obtain power by merely turning a faucet, as we now can gas or water . . .

In 1892 President Bartlett reached the mandatory retirement age of 75. Still vigorous and keen of mind, he nevertheless gladly gave up his heavy administrative responsibilities, which had never been easy for him largely because of the continuing opposition of some of the faculty, the alumni, and the Board of Trustees. He did not leave Hanover, but spent the next six years in the role of a successful and popular teacher at the College. He conducted a required senior course on the relation of the Bible to science and history until the time of his sudden death on November 16, 1898.

It was characteristic of the changes taking place in American education that the faculty of this period should have been very worried lest the new president of Dartmouth not be a professional educator, and nine members of the faculty, including C. F. Emerson, addressed a letter to the Trustees expressing this concern:

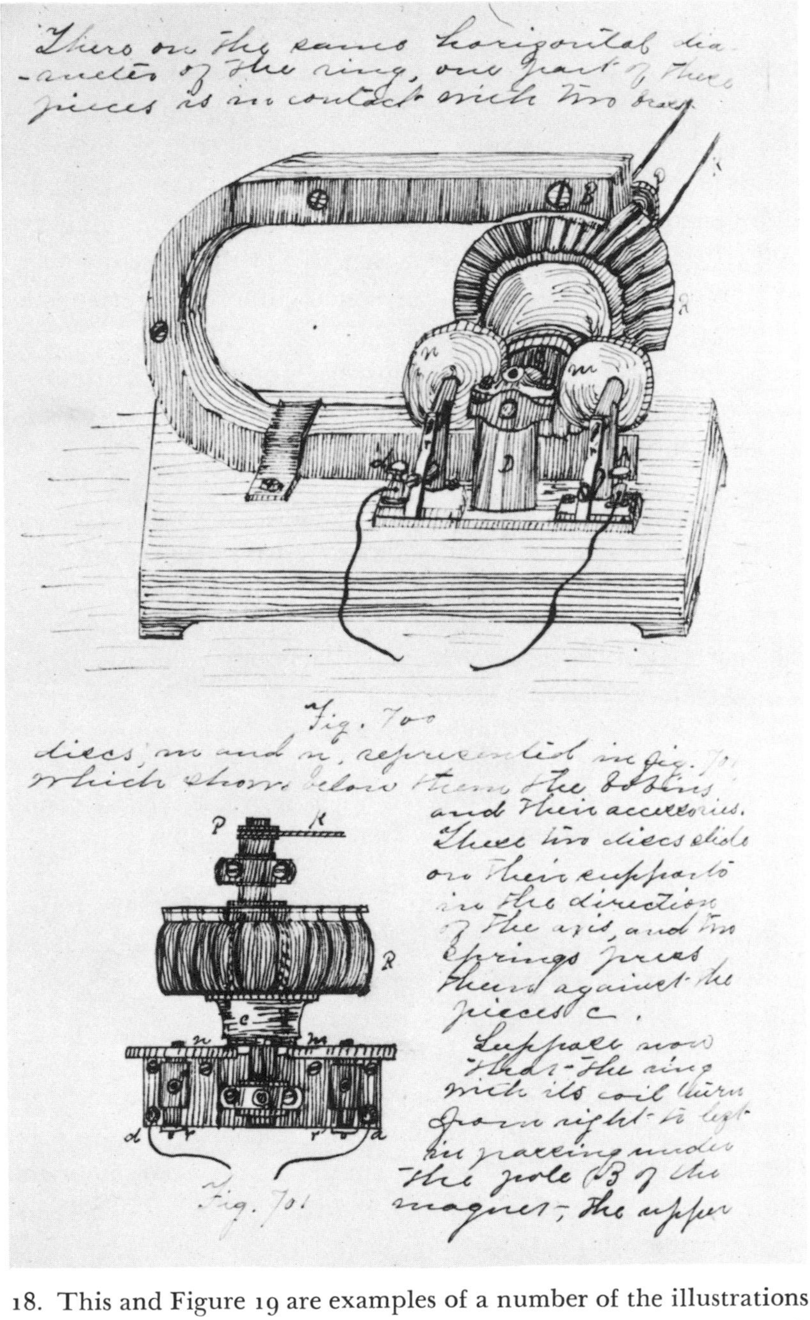

There on the same horizontal dia-
-meter of the ring, one part of these
pieces is in contact with two brass

Fig. 700

discs, m and n, represented in fig. 70,
which shows below them the bobbins
and their accessories. These two discs slide
on their supports
in the direction
of the axis, and two
springs press
them against the
pieces c.

Suppose now
that the ring
with its coil turn
from right to left
in passing under
the pole B of the
magnet, the upper

Fig. 701

18. This and Figure 19 are examples of a number of the illustrations which Mrs. Emerson made for the expanded physics textbook which Professor Emerson used in his classes. The special two-volume copy of Ganot's *Physics* was given by Mrs. Emerson to one of the authors of this book (S.C.B.).

by a lever *n*. the diaphragm, *b* which
is moveable in a slide. Above the

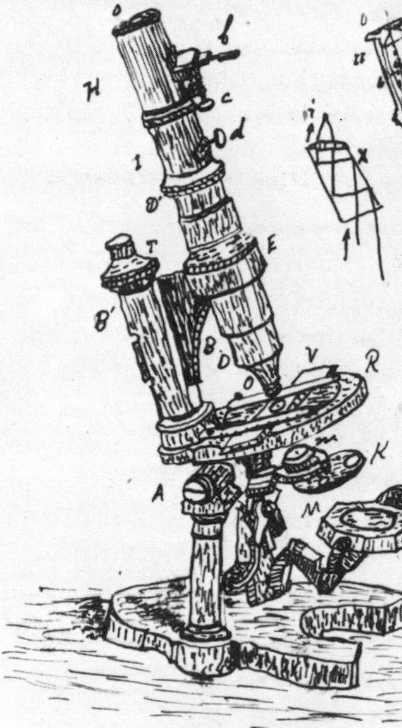

Fig. 481.

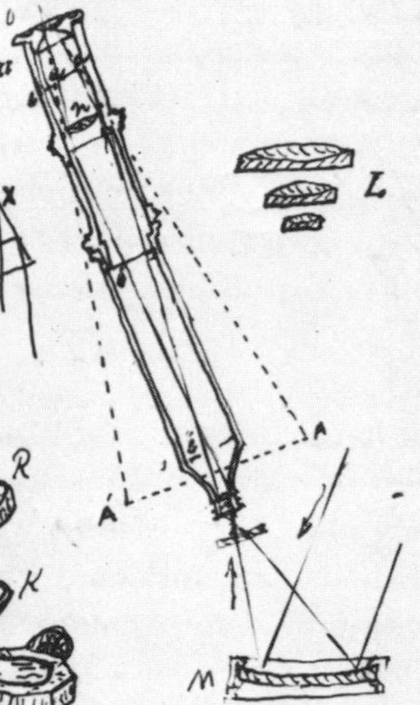

Fig. 432.

diaphragm is a piece, *m*. to which
can be attached either a very small
stop, so that only very little light
can reach the object or a condensing
lens, which illuminates it strongly,
or an oblique prism represented by *K*.
The rays from the reflector undergo
two total reflections in this prism,
and emerge by a lenticular face that
concentrates them on the object, but
in an oblique direction which in
some microscopic observations is an
advantage.

Our thoughts, perhaps naturally, turn more to the qualities which we hope to see in our new president than to his person. We remember that the College is a Christian College, and that upon the basis of worthy character it should seek to build the structure of generous scholarship. We expect to welcome a president who is both a Christian gentleman and a liberal scholar. But it is not enough, as it seems to us, that the person who comes to the presidency should have been merely a successful minister, lawyer, or man of business. Qualities which command success in any of these directions are desirable, but they are not enough. There is no sphere of activity which in the past few years has undergone such changes as education. On the one side it has become a science, on the other a profession, and it is not unreasonable to think that its chief positions can be adequately filled only by those who have had some experience in its methods and work, and whose practical acquaintance with its operation is not confined to their own training in public schools or to their college life. It is a suggested fact that in the choice of their last presidents, Harvard, Yale, Williams, Amherst and Brown did not select men because they were eminent in certain outside professions, but because they had had successful experience in collegiate training.[8]

To take President Bartlett's place, Dartmouth turned to one of its most beloved and well-known trustees, William Jewett Tucker, who at that time was professor in the Andover Theological Seminary. Dr. Tucker had a difficult time in deciding to leave his scholarly pursuits; but after a year of refusals he finally agreed to accept the position, it being obvious that Dartmouth needed him in a most desperate way. It must have given him a rewarding feeling to receive an outpouring of congratulations and welcomes, among which was a letter from Professor Emerson, who wrote, in part, "You will receive a most hearty welcome in Hanover and from all the professors."

One of President Tucker's first official moves was to reorganize completely the administrative structure of the college, relieving the President of many of his routine and time-consuming

8. March 31, 1892, addressed to Dr. C. P. Frost and held by the Archives of Dartmouth College.

responsibilities and creating permanent posts for duties which had been passed around the faculty in the past—and often accepted grudgingly. One of these changes was to create the office of Dean, to take charge of student records and day-by-day relations with the students. Professor Emerson was appointed to this position part time. At the same time his title was changed from Natural Philosophy to Physics, and a young assistant, Dr. A. C. Crehore, was appointed to help him in the teaching of physics. Dr. Crehore stayed at Dartmouth for seven years before going into industry and becoming one of the country's successful electrical engineers. Nominally, Professor Emerson would have held his position of Appleton Professor of Physics until 1899, but as time went on, more and more responsibilities were added to his already heavy list of commitments, and he turned an increasing amount of his work in physics over to his assistants. In 1898 his duties as a Dean were made full time, with complete relief from teaching duties.

Although in 1898 Emerson had given up all his teaching, he made what he considered his greatest contribution to the cause of science in the College by planning and overseeing the construction of Wilder Hall, providing the Physics Department for the first time with a building of its own. In his later years he often expressed a regret that he had never given a lecture or conducted a class in this building, which he worked so hard to create. The building was considered of such advanced design that a description of it was published in the leading American physical journal in 1901.[9]

Dean Emerson retired at the age of 70 in July, 1913, to spend nine happy years with his family and admirers. He was very proud of the fact that the records of the College showed that his forty-five years of continual service to Dartmouth was the longest on record. A number of pleasant tales are told of his retirement. One reception to him in Concord, New Hampshire, which attracted considerable notice in the press, involved his

9. The description of Wilder Hall is entitled "The Wilder Physical Laboratory of Dartmouth College," by E. F. Nichols, and is to be found in *The Physical Review,* Vol. 12 (1901), p. 366.

20. A portrait of Professor Emerson in middle life. As Professor John K. Lord wrote, "To his many qualities Mr. Emerson added good temper, a kindly manner, an active interest in many matters, and a fondness for sport that accompanied good health and vigorous physique."

21. The Wilder Physical Laboratory of Dartmouth College as it appeared in the year 1900.

22. It was from among this collection of tubes that Emerson, Frost, and Crehore found those particular ones which emitted X rays.

shaking hands with three of his former pupils one of whom was
the Governor of the State, one the President of its Senate, and
one the Speaker of its House.[10] Another story, coming from the
second trip he took abroad just after his retirement, was copied
from the Holyoke *Daily Transcript* of December, 1923, by one
of his former students:

> My interest in Professor Emerson and my sureness as to his
> memory dates back to April, 1914. On April 4, a little band of
> "Globe Trotters" found themselves visiting Pireas and Athens.
> We had group pictures taken on the Areopaus where Paul had
> made his remarkable statement as to the "Unknown God."
> While standing in the Pnyx in sight of Mars Hill and of the
> Acropolis, Professor Emerson recited in Greek a remarkable
> oration. With my eyes closed I might have imagined I was lis-
> tening to the shades of Demosthenes, Pericles, Sophocles, or
> Socrates, but when I asked Professor Emerson what he had
> been reciting, he replied, "my Greek Oration given in Dart-
> mouth when a junior."[11]

10. Reported in the *Concord* (New Hampshire) *Evening Patriot,*
April 7 and 8, 1921.

11. *Dartmouth Alumni Magazine,* Vol. 15 (1922–23), p. 486.

Chapter 9

A ONE MAN DEPARTMENT NO LONGER

A CHANGE was occurring in the concept of the college professor which was more than just a response to increasing student load or administrative responsibility. Professor Emerson had many years to observe the load which Professor C. A. Young tried to carry and which finally persuaded him to leave Dartmouth and go to Princeton, an issue that focused on the double duty of teaching and research. It was evident toward the latter part of the 19th century that the natural sciences were becoming a competitive research field and that excellence in teaching must somehow be geared to the rapid change in subject matter; thus it became incumbent on any first-rate teacher to keep up, at least in some measure, with the new knowledge by carrying out research of his own. It was in this era that for the first time in the history of Dartmouth the College had not a single professor of natural philosophy, but a professor of physics gathering around him a staff of young men who could share with him the teaching load, and also help him in his research projects. Woodward, Hubbard, Adams, Ira Young, Fairbanks, and C. A. Young were all in a one-man department; but within five years of the time that Emerson became Appleton Professor of Natural Philosophy, he persuaded the administration to give him some help, and from then until 1899, when he relinquished all his teaching duties, he was aided by a total of five men—Cook, Frost, Welch, Crehore, and McKee. None of these men stayed at the College permanently, but they introduced a kind of excitement which able young men can give to any institution, a circumstance totally lacking when a department has no new blood coming in, no young men trying teaching for the first time, as they push the department to allow them to try out their own ideas, unencumbered by the conservatism of older scholars. No attempt will be made to dwell on the careers of each of these men at any length, but they are worthy of more than a passing notice, because, for one important thing, their presence signals the eclipse of the natural philosophy concept by the growing professionalism and compartmentalization of physics. As noted, all of these first five assistants ultimately went elsewhere to make their marks as distinguished scientists in various fields. This pattern set by Emerson and his junior colleagues was by no means unique to Dartmouth. It was characteristic of the American educational scene, particularly in those fields of science that were bursting all bounds in their accelerated development. The trend has continued and increased in magnitude up to the present.

In the closing years of the 19th century, dramatic discoveries of the nature of matter were to render the physics of Woodward, Smith, Hubbard, and Young old fashioned and classical. It started in the *Sitzungsberichte der Würzburger Phyzik-medic. Gesellschaft* in 1895. The following is from the translation of an article, "On a New Kind of Rays":

> A discharge from a large induction coil is passed through a Hittorf's vacuum tube, or through a well-exhausted Crookes' or Lenard's tube. The tube is surrounded by a fairly close-fitting shield of black paper, it is then possible to see in a completely darkened room, that paper covered on one side with barium platinocyanide lights up with brilliant flourescence when brought into the neighborhood of the tube. . . . It is easy to show that the origin of the flourescence lies within the vacuum tube.
>
> It is seen, therefore, that some agent is capable of penetrating black cardboard which is quite opaque to ultraviolet light, sunlight or arc-light. It is therefore of interest to investigate how far other bodies can be penetrated by this same agent. It

is readily shown that all bodies possess this same transparency, but in varying degrees. . . . Thick blocks of wood are still transparent. Boards of pine two or three centimetres thick absorb only very little. A piece of sheet aluminum 15 mm. thick, still allows the X-Rays (as I will call the rays for the sake of brevity) to pass, but greatly reduced the flourescence. . . .

Of special interest . . . is the fact that photographic dry plates are sensitive to X-Rays. It is thus possible to exhibit the phenomena so as to exclude the danger of error. . . . The photographic plate can be exposed to the action without removal of the shutter of the dark slide or other protective case, so that the experiment need not be conducted in darkness. . . .

The retina of the eye is quite insensitive to these rays; the eye placed close to the apparatus sees nothing. It is clear from the experiments that this is not due to want of permeability on the part of the structures of the eye. . . .

I have also a shadow of the bones of the hand; or a wire wound upon a bobbin; of a set of weights in a box of a compass card and needle completely enclosed in a metal case; of a piece of metal where the X-Rays show the want of homogeneity, and of other things.

The paper is signed by W. K. Röntgen.[1]

The issue of *Science* for February 14, 1896, contains reports from three American laboratories where physicists had figured out what Röntgen had been doing. The earliest of these, from Hanover, New Hampshire, dated February 4, 1896, begins:

Experiments with Röntgen's newly discovered X-Rays have been carried on during the past few days in the Dartmouth physical laboratory by Professor C. F. Emerson and the writer, and some of the preliminary results already obtained may be worth recording.

Of four Crookes tubes first tried, but one emitted rays which (with the exposure given) made a visible impression upon a

1. Röntgen's original paper was made available in this country through a translation by Arthur Stanton in *Nature* (January 23, 1896, page 274) although there was a brief note about it in the preceding issue of *Nature,* ending with "The scientific world will look forward with interest to the publication of the details of Prof. Röntgen's work."

photographic plate protected from the ordinary luminous rays. This tube is 4.7 cm. in diameter and is cylindrical for a length of 16 cm., then tapering to a point. The platinum electrodes are in opposite sides of the cylindrical surfaces and are about 5 cm. apart. The phosphorescent plate is interposed obliquely between the electrodes. In action the phosphorescent surface is bombarded by the discharge from the negative pole. We have thus far usually excited the tube by a current from an efficient induction coil, but a Holtz machine has served about equally well.

The first successful experiment gave, after 12 minutes of exposure, a picture of a knife and scissors hung on the side (1 cm. thick) of a white wood box, within which the photographic plate has been placed.

Subsequently the Crookes tube was supported horizontally, and the plate holder could then be laid upon the table and any object interposed that was desired. No camera was employed, and the slide of the plate holder was not drawn, so that no exposure to the ordinary liminous rays could occur.

A coin and key concealed between two boards of total thickness 24 mm., were shown after an exposure of 11 minutes, the tube being 15 cm. above the plate. . . . Silver and gold seem to be the most opaque of the metals yet tried, although aluminum transmits poorly. Glass is more opaque than brass, and less so than hard rubber. Cork transmits better than any other substance examined. . . .

It was possible yesterday to test the method upon a broken arm. After an exposure of 20 minutes the plate on development showed the fracture in the ulna very distinctly. Comment upon the numerous applications of the new method in the sciences and arts would be superfluous. Edwin B. Frost.

Continuation of this exciting scientific chase was reported again in the issue of *Science* of March 27, 1896, under the title, "Further Experiments with X-Rays":

Photographs have now been obtained with several of the Crookes tubes in the cabinet of the Dartmouth Laboratory, but the one referred to in a previous communication is by far the most efficient, and it has been used in nearly all the experiments now to be described. The tube was made by Stoehrer of Leipzig, being No. 1147 of his catalog, where it is designated as Puluj's neue Phosphorescenz-Lampe. It contains a mica diaphragm coated with some phosphorescent substance, and gives quite a brilliant green light when in action (although this brilliancy is doubtless immaterial to the production of the X-Rays).

As to the source of the X-Rays developed by this tube it may be stated that a variety of experiments have shown that they originate in the diaphragm itself where exposed to the cathode rays and not to any appreciable degree in the glass around the diaphragm. Cathode rays which pass through the diaphragm appear, however, to develop X-Rays at the surface of the glass where they impinge. . . .

The difficulty found by Professor Emerson and myself in *precisely repeating* most of the experiments has doubtless been experienced by others working with the X-Rays. When the conditions of an exposure seem identical with those of a previous one, the results often differ, from varying excitation of the tube, or possibly slight shifting of the source of rays, or from numerous other causes difficult to control. A confirmation of results by other observers is therefore valuable. . . . Edwin B. Frost, Hanover, New Hampshire, March 10, 1896.

Having thus introduced the most dramatic experiments that have ever come out of the laboratories of physics at Dartmouth College, we go back to be introduced to the energetic young man who was the writer of these papers in *Science*. But before interrupting the X-ray story completely, one might quote briefly from Frost's own recollection of these dramatic events:

During the years 1887–89 the writer had been assistant to Professor Emerson in the Physical Laboratory on the ground floor of Reed Hall, and had the privilege of using the apparatus there. When the cable hints were received about Roentgen's success, it immediately seemed worthwhile to test the numerous vacuum tubes in our laboratory for their capacity to produce the mysterious rays. Our good friend H. H. Langill, the local photographer, who was always glad to assist in scientific experiments, took care of the developing and printing of the pictures. The results were immediate and startling, particularly since the Roentgen article in *Nature* had not then arrived. . . . None could ever forget the interest felt in watching the development

of those first plates on that Saturday evening either January 24
or February 1, 1896, probably the latter. It did not take long to
find where the rays were most active around the tubes, and the
Puluj tube proved to be the most efficient. I suspect, in fact,
that it was one of the best tubes in America for the next few
weeks.[2]

Edwin Brant Frost was clearly one of those instrumental in
ending the old era and he interacted with a number of the other
figures represented in this book in a very positive way. Although not born in Hanover, he considered himself a local boy,
having been brought here at the age of 5 when his father (Dr.
Carlton P. Frost) accepted a professorship in medicine at Dartmouth in 1871. Edwin Frost, who subsequently became one of
the leading astronomers of the world and director of the Yerkes
Observatory, wrote a charming autobiography, in which he
detailed many of the homey events of his growing up.[3] He re-

---

2. Frost's recollections of the first X-ray experiments in America are
contained in an article he wrote for the *Dartmouth Alumni Magazine.*
April 1930.

3. Frost's autobiography is entitled *An Astronomer's Life,* and was
published by Houghton Mifflin Company, Boston, 1933. It is a book well
worth reading for a very personal account of the life of one of our most
famous astronomers. The general public was very much interested in Frost,
not only because of his scientific reputation and his warm personality, but
the fact that he went blind and still was able to take visitors to the observatory and point the telescope to the proper places in the heavens to identify
the sights, and to give public lectures without the audience always being
aware that he was blind. Tilton Boutin in the May 1943 issue of *Popular
Astronomy* tells of a few of these instances:

Once after Professor Frost had returned having been absent about
seven months, I went immediately to call upon him: He was sitting
alone on a piazza. I approached and said "how do you do Dr. Frost."
Instantly came his reply "Why, Mr. Boutin. I'm glad to see you."
Then after talking a few minutes he told me that just before I arrived
a lady reporter from one of the newspapers had interviewed him, and
he added with a smile "I don't think she suspected that I could not
see her."
One bright evening he came with Mrs. Frost and some friends to
the observatory. . . . The roof of the observatory rolls entirely off
giving a full view of the sky. Professor Frost felt along the polar
axis of the mounting, stepped directly back, and I saw him point to

calls, for example, that "we boys detested those little velvet suits given to us by the Fairbanks family after several children had outgrown them. We were supposed to look so elegant, however, that we had to be photographed in them."

The Frost family and the Young family were very close to each other, and their children grew up playing in each other's houses. Some indication of Edwin's early interest in science is provided by a notebook in the Dartmouth Archives recording "Temperatures taken at about 7 A.M. and 10 P.M. by Fahrenheit thermometer—From North Window second story of Dr. C. P. Frost's house Wheelock St." for the years 1882–87. It was Young who, probably more by example than by persuasion, turned Edwin Frost into an astronomer. Frost's own words are interesting:

> Senior year at Dartmouth in 1886 provided rather a large range of electives. I had been greatly interested in a course in descriptive astronomy given by Professor Charles F. Emerson (familiarly known as "Chuck"), and had up to then used the telescope, with which I had been more or less familiar for many years. . . . I was considerably in doubt as to whether my work would be in the line of physics or astronomy. The tradition of Professor Young's brilliant work in the observatory was still strong. My personal acquaintance with him and the family added to the interest.
>
> In the summer of 1888, Professor Young was working intensively upon his series of textbooks, the first and largest of which was the *General Astronomy* bearing the imprint of 1888 and published by Ginn and Company. Professor Young asked

---

the stellar position as accurately as he could have done had he been able to see. The heavens were in his mind.

This remarkable ability to hold in mind a mental picture was equally shown when he lectured. He could accurately announce the positions on the screen of any object of which he was speaking. Just once I saw him hesitate. He stated the location of an object then paused a moment and asked Mrs. Frost, whose custom was to sit near him on the platform while he lectured, "Is that correct." A gentleman touched me on the shoulder and whispered "Why does he ask her? Can't he see for himself?" When I whispered in reply "He is totally blind" the man was astonished."

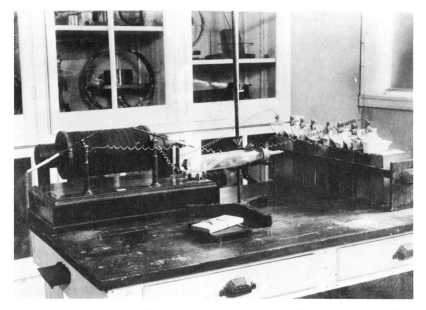

23. A contemporary picture of the Dartmouth Crookes tube, which turned out to be such an excellent source of X rays. X rays are emitted when a beam of electrons is suddenly stopped, and the X rays are emitted at right angles to the direction of travel of the electron beam. As can be seen in this picture, the mica screen, making an angle of 45° with the electron beam, was by chance optimumly situated to produce copious X-ray emission when the pressure in the tube, which was also at a fortunate value, was within a reasonable range for the voltage available. This picture also shows the coin and the key with the small wooden board described in the report to *Science*.

24. The Nichols and Hull Experiment set up in Wilder Hall.

me to read the proof, which arrived about twice a week and kept me pretty busy. It was my first experience in proof-reading, and I wished to solve all the problems used for exercises in order to check their accuracy. My small assistance to the publisher and author was of course of far less consequence to them than the value of the experience to me. The book was a great success, being widely used in colleges.

In showing how much Frost followed in the footsteps of C. A. Young, it is also worth pointing out that during the total eclipse of the sun of May 28, 1900, both Young and Frost were at their observing stations at Wadesboro, North Carolina, during the totality, and as Frost points out, "It was my part to obtain photographs of the so-called 'flash spectrum,' at the beginning and end of totality; also to secure if possible photographs of the spectrum of the corona." And at the death of Professor C. A. Young, it was E. B. Frost who was asked both by the National Academy and the Astrophysical Society to prepare biographical memoirs in Young's honor.

Professor Frost had a congenital defect in his eyes which eventually led to blindness, but at the time he was at Dartmouth this gave him a great advantage. He was very near-sighted. By removing his glasses and holding the photographic plate an inch from one eye, he could bring it to a focus on the retina and thus had a magnifying power of ten diameters without using a microscope. By this means he could scan plates with remarkable speed, and he became the world's chief specialist at detecting binary stars.[4]

As a footnote to the fact that Professor Emerson felt his greatest contribution to science at Dartmouth had been the design and construction of Wilder Laboratory, it may be noted that E. B. Frost reveals that "the Wilder gift was obtained chiefly by the suggestion of my father, who had friendly relations with Mr. Wilder, both personally and as his physician."[5]

---

4. The comments about Professor Frost's capitalization of his eye defects come from a tribute to him on his death by Gordon F. Hull, C. A. Proctor, and Richard H. Goddard. A copy is in the Dartmouth Archives.

5. We have only E. B. Frost's comment on p. 92 of his autobiography as authority for the statement that Dr. Carleton Frost persuaded Wilder to

(The elder Frost, Dr. Carleton P., was Dean of the Medical College, a most successful teacher, and had a large local practice. He served for some years on the Board of Trustees of the College, and he was obviously well aware of the necessity for improving the physical plant of the Department of Physics.)

As has been mentioned before, when Professor Emerson was appointed a part-time dean, Dr. A. C. Crehore came to Dartmouth, mainly to assist in the teaching of physics. Crehore was more of an inventor and an engineer than he was a physicist, but it is an interesting comment on the universal enthusiasm which followed Röntgen's discovery of X rays that Crehore himself remembers rushing into the laboratory and looking for Crookes tubes which could be used for X-ray sources: "The discovery of X-Rays in 1895 was announced while I was at Dartmouth. Going to the collections of good Crook's vacuum tubes already in the laboratory, my student, Mr. F. E. Austin, who later became professor in the Thayer School of Engineering in Hanover, made a clear picture of the bones of his hand. How easy it would have been for me to have done this years before. All the means were at hand, but we had to be shown how."[6]

Crehore came to Dartmouth with a fresh Ph.D. from Cornell, having just co-authored a book on alternating currents which became a standard text on this subject for some years thereafter.[7] It is of some interest that Ernest Fox Nichols, the subject of the next chapter, was a student at Cornell at the same time as Crehore, and they boarded at the same place.

Dartmouth was proud of its physics apparatus. A section in the Dartmouth College Catalogue of 1890–95, reads in part:

_____
give this gift to the College. C. P. Frost died a year before the Wilder gift was received, and there is no record in the deed of gifts to suggest how Wilder was persuaded to be interested in the physics department.

6. Many of the details about Crehore came from a small book entitled *Autobiography*, Albert Cushing Crehore, Ph.D., published by William G. Berner, Gates Mills, Ohio, 1944.

7. *Alternating Currents: An Analytical and Graphical Treatment for Students and Engineers,* by Frederick Bedell and Albert Cushing Crehore, published by the W. J. Johnston Company, Ltd., Park Row, New York, 1893.

"The Philosophical apparatus of the Appleton Professorship is very extensive and valuable, and sufficient for all the purposes of illustration. In the departments of acoustics, optics, and electricity, it is especially rich. Two additional rooms have recently been fitted up for physical laboratories."

Using the equipment in optics, Dr. Crehore was able to devise what he called a "Polarizing Photo-Chronograph," which involved the excellent Nicol prism polarizers, still present in the College's laboratories, and the Faraday effect in carbon bisulphide. This created a rapidly operating optical shutter, and he initiated a long series of experiments to demonstrate the use of this device for measuring the velocity of projectiles, publishing his results in the *Journal of the U.S. Artillery*. He also received two patents, which he sold to the General Electric Company, relating to electric railroads. The patents involved a method of making contact with the electric power by means of contact buttons along the railroad track which are dead except for the brief time when the car is directly over them. He invented, as well, a method of sending telegraph messages at very high speeds by using his polarizing photochronograph on alternating current power sources.[8] This device was not unlike our modern magnetic storage disks, except that the information was stored on paper disks saturated with a chemical that recorded as a steel needle carrying current was drawn across it. He later recalled "sending by means of a perforated tape the whole of Longfellow's poem, 'The Village Blacksmith,' and recording it upon this chemical receiver in nine seconds. . . . The first long test of this receiver that I recall was made in a laboratory in Washington, D.C. This remains a memory because Dr Alexander Graham Bell was conducting some kind of experiment in another part of the large room. I met him at this time and afterwards attended one of the seminars that he was accustomed to hold at his residence. At one of these I met Professor George F.

---

8. Everybody at the College seemed very much interested in Professor Crehore's "invention for the rapid transmission of intelligence by electricity," and there were two long articles in *The Dartmouth* in the issues of 1896–97 and 1898–99 about his "synchronograph."

Barker of the University of Pennsylvania, whose textbook in
general physics I had been using for physics classes in Dartmouth."

So intrigued was he by this device that he started a company
known as The Crehore-Squier Intelligence Transmission Company, and during the summer of 1897 he went to London to
offer this invention to the British government, demonstrating it
on a line between London and Aberdeen, Scotland. Soon after
the company was incorporated he left Dartmouth and moved to
Cleveland, to spend his full time designing telegraph apparatus.

Chapter 10

THE END OF AN ERA

IN 1898, at the age of 29, Ernest Fox Nichols accepted appointment as Professor of Physics at Dartmouth College, after receiving the Doctor of Science degree from Cornell. He was to stay five years, then leave, and finally return in 1909 as President of the College, following Dr. Tucker in the "Wheelock Succession." In his brief tenure on the faculty he initiated an era of commitment to scientific research in physics which brought to the institution an international reputation.

Nichols was born on June 1, 1869, in Leavenworth, Kansas. Early biographical notes from an article by his cousin, Philip Fox reveal that

> The youth Nichols did not have the usual routine of school experience, for he was rather delicate. His mother died before his 13th birthday and he employed himself with his older brother Arthur distributing newspapers. Two years later his father died, leaving the two boys alone. The household was broken up and Ernest went to Manhattan, Kansas, to the home of his uncle, General S. M. Fox. There he entered Kansas State Agricultural College at the age of 15, a youth without a day of formal schooling. He was allowed to enter college on the sole condition that he make good, and with the support of President Fairchild, to whom his case had strong appeal.[1]

---

1. *The Astrophysical Journal*, Vol. 61, No. 1, January 1925.

Nichols' aunt, Mrs. S. M. Fox, is quoted in this same article:

> In many ways he was much better informed than most boys his age; his mother was well-educated, an excellent conversationalist, and his father was strong in chemistry and astronomy and inclined to research. In other things a bright boy in the fourth grade was ahead of him—such writing, such spelling. He told me when he came here that he had read only two books through in his life. But his mother had read to the boys a great deal of the best sort. Both of the boys spoke correct English because they had never heard anything else. The scientific bent was from his father wholly.

He finished his four years, receiving the degree of Bachelor of Science among the three or four strongest students. Obviously his aunt, his uncle, and his brother had encouraged him all the way.

He stayed on at Manhattan for an additional year of work, acting as an assistant in chemistry. He also looked after the meteorological instruments and records. At this time, too, he kept a daily record of solar activity, mapping the position of sunspots by projecting the solar image upon a screen.

In the fall of 1889 he entered Cornell University at the age of 20 to begin graduate work. Here he constructed a galvanometer that was one of the most sensitive of its kind ever made, to be used in conjunction with a thermopile for absorption measurements of infrared radiation in various optical media. The galvanometer was so sensitive that artificial disturbances in the earth's magnetic field or auroral disturbances at night would drive the spot of light across the scale. The faintest gleam of northern lights was the signal for a postponement.[2]

In 1892 he received the degree of Master of Science from Cornell, and in the fall accepted the position at Colgate University as Professor of Physics and Astronomy. He went to Berlin in 1894, shortly after his marriage to Miss Katherine West,

---

2. The description of his work at Cornell taken from E. L. Nichols, *Biographical Memoir of Ernest Fox Nichols*, the National Academy of Sciences, *Biographical Memoirs*, Vol. 12, Sixth Memoir (1928), p. 101.

the daughter of one of the foremost citizens of Hamilton. There he developed a refined radiometer and published the classical paper "A Method for Energy Measurements in the Infrared Spectrum and the Properties of the Ordinary Ray in Quartz for Waves of Great Wave Length."[3] His very sensitive radiometer, employing a light suspension fiber of fused quartz, operated in a gas pressure of 0.05 millimeters of mercury and was sufficiently sensitive that a candle at a distance of 1 meter caused a deflection of 2100 scale divisions.

By the time he had published several papers with Rubens in Berlin, Nichols at the age of 28 was regarded both in Europe and America as an experimental physicist of extraordinary ability. Upon his return to Colgate in 1897 he presented three papers to the faculty at Cornell University for the degree of Doctor of Science, and in the following year he was called to Dartmouth as Professor of Physics.[4] Shortly thereafter he was designated Director of the Wilder Laboratory, to which the department moved from Reed Hall in the fall of 1899.

The laboratory was built in the age of electricity and spectroscopy. The building was rich in laboratories and not overly burdened with lecture rooms or recitation rooms. It was truly designed for scientific work. In addition to an electrical laboratory and magnetic laboratory, spectroscope, concave grating, photographic laboratory, photometric laboratory, an optics laboratory and acoustics, there were supporting facilities including dynamo room, battery room, equitemperature room, an electrical room, many photographic darkrooms, chemical kitchen, several laboratories for instruction, well-placed rooms for apparatus, and a library. There were carefully thought-out cir-

---

3. *Physical Review*, Vol. 4 (1897), p. 297, from biographical memoir by E. L. Nichols.

4. Nichols' thesis was entitled "Radiometric Researches in the Remote Infra-Red Spectrum" and was published in 1897 after his return to Colgate. The three papers comprising the thesis were "A Method for Energy Measurements in the Infra-Red Spectrum and the Properties of the Ordinary Rays in Quartz for Waves of Great Length"; "Heat Rays of Great Wave Length," jointly with Heinrich Rubens; and "Certain Optical and Electro-Magnetic Properties of Heat Waves of Great Wave Length," also with Rubens.

cuits for carrying electrical signals anywhere in the building,
particularly in connection with a clock and chronograph. In
addition, arrangements for mounting a heliostat that would
project a sunbeam along the lecture table was typical of the
thoughtfulness that went into the planning.

The timing of construction makes it evident that Professor
Emerson was the person responsible for the interior designs,
although elsewhere it is evident that Dr. Nichols influenced
these decisions once he had arrived in Hanover. Clearly this
facility was designed for research in physics by the faculty in
conjunction with their students. It was self-contained, not dependent on other departments or resources, even to the point of
including chambers and studies for the instructors who in every
sense were full-time in the Department of Physics. Later, after
Nichols' departure, Wilder became increasingly a teaching laboratory and the faculty primarily a teaching faculty. A number
of laboratories were turned into recitation rooms to accommo-
date the large number of students required to take physics or
electing astronomy. These changes resulted from a deliberate
change in policy away from an emphasis on research toward a
department almost wholly occupied by its teaching function.

Nichols was at Dartmouth as Professor of Physics for only
five years. Throughout this period he worked with an intensity
characteristic of a contemporary physicist, which, combed with
his judicious selection of fundamental experiments, led to striking accomplishments in so few years. He was early invited by
the Yerkes Observatory to prepare a special radiometer designed
specifically for stellar observation.[5] The Nichols Radiometer
was sufficiently sensitive to detect changes in the number of
candles in a group at a distance of 16 miles. Through such experiments, he was able to study the absorbing effect of the atmosphere between star and observer, and thus correct his measurement of the energy output of stars and planets.

The experimental work on the pressure of light which is

5. Radiation intensity from stars and planets was published by Nichols as a paper: "On the Heat Radiation of Arcturus, Vega, Jupiter, and
Saturn," *Astrophysical Journal,* Vol. 13 (1901), p. 101.

most often associated with Nichols' role as a physicist at Dartmouth, and indeed along with the work of Professor C. A. Young, represented the establishment at Dartmouth of the modern tradition in physical research. This work was begun in the Wilder Laboratory with the collaboration of G. F. Hull, then at Colby University, whom Nichols had invited to Dartmouth as Assistant Professor in January of 1900. Hull had shortly before completed his Ph.D. degree under A. A. Michelson at the University of Chicago, and Nichols specifically sought as a collaborator a young experimentalist whose presence would complement his own work in the field of radiation. The experiment carried out by them was a measurement of the pressure of light which had been mathematically predicted by Maxwell in 1873. One can get a sense of the pulse of science then as compared with now in that twenty-five years elapsed between the articulation of such a striking phenomenon and its detection and verification. There is some evidence[6] that Nichols had had such a measurement in mind even during his stay in Berlin, following the development of the very sensitive radiometer. The motivation for the measurements was the result derived by Maxwell that "light consists in the transverse undulations of the same medium which is the cause of electric and magnetic phenomena" and that therefore it has an electromagnetic momentum in free space which should be measurable as an ordinary momentum. Light should exert a minute pressure in falling on any surface.

For a quarter of a century attempts at measuring this phenomenon came to nought, both because of a lack of sufficiently sensitive suspensions to detect the slight pressure and because of the overwhelming effects of gas actions vital to the operation of the conventional radiometer. As Nichols and Hull wrote:

> Every approach to the experimental solution of the problem [the detection and measurement of radiation pressure] has hitherto been balked by the disturbing action of gases, which it is impossible to remove entirely from the space surrounding

---

6. E. L. Nichols' biographical memoir, p. 109, makes the suggestion that Nichols had earlier contemplated this experiment.

the body on which the radiation falls. The forces of attraction or repulsion due to the action of gas molecules are functions of the temperature difference between the body and its surroundings, caused by the absorption by the body of a portion of the rays which fall upon it and of a pressure of the gas surrounding the illuminated body. Indeed, such was the basis of the Nichols radiometer.[7]

It was Nichols' confidence in his ability to refine his sensitive radiometer which led him to undertake this crucial experiment. Conceptually the measurements were rather elementary. They required the calibration of the torsion balance, so that rotation of a tiny paddle wheel suspended from a thin, quartz fiber could be related to the force resulting from the pressure of radiation from an arc lamp. The genius of the experimenters was their approach to the reduction of the disturbing effect of gases in the vessel housing the torsion system. To this task they held with great tenacity throughout the four-year period of experimentation. The initial results of their work were first presented to the scientific community in a preliminary paper read at the Denver meeting of the American Association for the Advancement of Science in August 1901.[8] The experiment was completed in February 1903, and the definitive work was published in *The Physical Review* in July and August 1903.

---

7. "A Preliminary Communication on the Pressure of Heat and Light Radiation," by E. F. Nichols and G. F. Hull, *Physical Review,* Vol. 13 (1901), p. 307.

8. As was so often the case in those times, similar experiments were carried out in different parts of the world, unbeknownst to the individual experimenters. So in the measurements of radiation pressure, Levedew in Moscow was conducting similar measurements. His measurements initially published in *Annalen der Physik,* Vol. 6 (1901), p. 433, appeared as an abstract in the *Astrophysical Journal* for January 1902, which also contained an abstract of the paper delivered by Nichols and Hull in Denver, August 29, 1901. The first complete reports were published in *The Physical Review* by Nichols and Hull and in *Annalen der Physik* by Lebedew, appearing in the respective November 1901 issues of those two journals. Actually, Lebedew made a preliminary report at the International Congress of Physics, which met in Paris in the summer of 1900. The accuracy of his measurements never approached that of Hull and Nichols.

The environment around Wilder Laboratory and the role of Nichols is best described by Professor Hull:

> The years 1899–1903 were strenuous and exhilarating years for the Department of Physics. New equipment was constantly being added, new courses were being offered, and very exacting research was being carried on. Vacations, except for tramping trips in the mountains, were largely given up to research. In all of this work Dr. Nichols was the invigorating spirit. The successful completion of certain research work, notably that having to do with the detection and measurement of the pressure of light, together with his breadth as a teacher, gave Dr. Nicols a prominent place in the educational world. Naturally he was wanted elsewhere. and in 1903 he was offered and he accepted a professorship of physics at Columbia University. Before he severed his connection with Dartmouth, however, he was given the honorary degree of Doctor of Science in recognition of the significant service he had rendered to science and to Dartmouth.[9]

In his five-year period at Dartmouth Professor Nichols must have concentrated his interests in his teaching and research in Wilder Laboratory. Yet following the announcement by President Tucker in 1909 that he intended to resign, Nichols was called to the Presidency. One can only conclude that he had left a profound impression upon the college community of Hanover as a man of character and ability. After six years at Columbia, Nichols accepted this call back to Dartmouth at an obvious scientific sacrifice. Many of his friends took exception to his decision,[10] one of whom suggested that "there were a thousand men qualified to do fairly well what a college president had to do. Where will you find another to measure the pressure of light or determine the heat from the fixed stars? . . . if by some cataclysm Dartmouth College . . . were to be swept from the face of the earth, it would be utterly forgotten in a single century; say

---

9. Professor Hull was writing in the *Dartmouth Alumni Magazine,* June, 1924, after the death of Dr. Nichols.

10. The observation suggesting that Nichols was being taken from his scientific work was made by an unidentified speaker at the inaugural dinner and is quoted by E. L. Nichols on p. 115 of the biographical memoir.

25. Ernest Fox Nichols with one of the architects of the new physics, Albert Einstein.

that it might be recalled that here in the opening years of the 20th century Nichols and Hull carried on their imperishable researches." And as David Fairchild put it, "I cannot help feeling that my dear old friend and classmate was a victim of the mistaken idea that it is more important to have a good college president than to have a good research man. No matter if Ted was a good president—he was a real investigator and there are ten good presidents—yes, twenty—to one really great investigator."

Although his friends may have differed in their estimate of the relative worth and availability of college presidents as opposed to research men of genius, there was a widely held opinion that Dartmouth had encroached on Dr. Nichols' physics. He himself, as it proved, did not really find administration to his liking, but soon returned to research and physics. In many ways he was the model of the physicist who, with single-minded purpose, follows his scientific research as his guiding star. And he died as he lived. While presenting a paper, "Joining the Infrared and Electric Wave Spectra," before his peers at the National Academy of Sciences in Washington on April 29, 1924, so long a pause occurred in his delivery that the chairman finally walked to the podium to find Nichols had died of a heart attack while remaining standing before them, propped up by the dais.

EPILOGUE

THE reader who has come to the end of these sketches should be struck by the ups and downs of excellence and enthusiasm as one man followed another in their professional roles. It was characteristic that when men were willing and able to carry on vigorous research, the college's fame as a leading educational institution was greatly enhanced, and when the main concentration was just on meeting classes and performing the day-to-day duties of heavily loaded teachers, the education tended to be pedantic and ordinary.

When there was but a single man to carry the burden of a subject, the time constant of the oscillation between leadership and mediocrity was a single teaching generation, but as the college grew, so did the inertia for change, and the ups and downs were apparent much more slowly.

Coincident with Dartmouth's inability to hold such national leaders as Nichols and Young, rose the popularity of an educational doctrine that declared a dichotomy between teaching and research. This was by no means just at Dartmouth, but it became an accepted conclusion all over the country that so rare was it to find excellence in teaching and research combined in a single man that "liberal arts colleges" went out of their way to find the "dedicated teacher," and if he wanted to do some research in the summer or on his own time, that was his own business; it was not encouraged by the administration as part of his real commitment to the college.

This occurred dramatically at Dartmouth. The great open research laboratory spaces designed with such pride by Emerson and Nichols were closed in and filled with desks and black-boards. Into Wilder Hall moved other departments for their classrooms and offices—mathematics and astronomy shared space with physics, and within a decade or two after it was built, Wilder Laboratory was almost indistinguishable from any other classroom building on the campus.

The College grew and so did the inertia for change, but the country's educational philosophy changed also. The concept that a great teacher must concentrate exclusively on his teaching was replaced by an almost complete reversal of outlook as more and more professors all over the country demonstrated that their inspiration as great teachers was tied to a combination of teaching and research. Dartmouth recognized this change and started back on the upswing to national prominence of educational excellence, but the inertia of a larger institution makes changes come more slowly and the maxima and minima less pronounced. Educational leadership is no longer so dependent on the abilities of an individual man but rests much more with the whole faculty. Nevertheless, the simpler model of the smaller institution may still be useful and hopefully may provide insight and guidance in making today's changes in educational philosophy toward the better.

INDEX

Adams, Ebenezer, 34–43, 44, 45, 46, 68, 101
Adams, Ebenezer, Jr., 35
Adams, Eliza, 46
Allen, William, 40
Alumni, 65
Amherst College, 54, 96
Andover Theological Seminary, 59, 68, 96
Appleton, Samuel, 47
Austin, F. E., 110

Baldwin, Loammi, 23
Bartlett, Samuel, 78, 88, 89, 93, 96
Bissell Gymnasium, 86
Bridges, Robert, 84
Brown, Francis, 40, 42
Brown University, 96
Burlington, Iowa, 71, 72, 73, 87

Chandler Scientific School, 53, 89
Cleveland, Ohio, 69
Columbia University, 119
Cook, Charles Sumner, 101
Crehore, Albert C., 97, 101, 110, 111, 112
Crookes tube, 108

Dana, Daniel, 42
Dartmouth, Lord, 12

Dartmouth College Case, 38
Dartmouth Hall, 4, 14
Dartmouth Scientific Association, 86, 92
Dartmouth University, 11, 40
Dean, James, 40
Deerfield Academy, 26

Emerson, Charles F., 85–100, 101, 102, 103, 105, 107, 116, 123
Emerson, Mrs. Charles F., 91

Fairbanks, E & T, & Company, 58, 60
Fairbanks, Erastus, 58
Fairbanks, Henry, 57–67, 68, 70, 101
Fairbanks, Thaddeus, 58
Fairchild, David, 121
Flash spectrum, 76
Frost, Carleton P., 106, 107, 110
Frost, Edwin B., 101, 105, 106–109
Frost, Mrs. Edwin B., 106, 107

Haddock, Charles B., 46
Handel Society, 31
Hanover, New Hampshire, 39
Hartford, Vermont, 39
Harvard University, 17, 96
Hitchcock, Hiram, 65

Hubbard, John, 14, 26–33, 34, 40, 44, 45, 101, 102
Hubbard, Oliver P., 54
Hull, Gordon F., 117, 118, 119

Kimball, Samuel A., 5, 6, 29

Langill, H. H., 105
Lebanon, New Hampshire, 45
Leicester Academy, 34
Light, pressure of, 108
Lord, John K., 89
Lord, Nathan, 44, 45, 49, 56, 57, 59, 60
Lyndon Academy, 58

Martha's Vineyard Summer Institute, 90
Mary Hitchcock Memorial Hospital, 65
McKee, George C., 101
Meriden Academy, 46
Moor's Charity School, 2
Middlesex Musical Society, 27
Mixer, Augusta, 69
Mount Young, 82

New Hampshire College of Agriculture and the Mechanic Arts, 86, 89
New Ipswich Academy, 14, 85
Nichols, Ernest F., 97, 110, 113–121, 122, 123
Noyes, Annie, 62
Noyes, Daniel J., 62

Orrery, 36

Parish, Elisha, 27
Patterson, James, 60, 63, 64
Peabody, Asa, 5
Peking, 76
Phillips Academy, Andover, 68
Phillips, John, 12
Portland Academy, 34
Powell, James, 62
Princeton University, 2, 77, 79

Quimby, Elihu T., 85

Reed Hall, 47, 57, 92, 105, 115
Ripley, Sylvanus, 16, 37
Rose, William, 14

St. Johnsbury Academy, 58
Sanborn, Edwin D., 59
Semi-circumferentor, 6, 10
Shattuck, George C., 49, 51, 54
Shattuck Observatory, 50, 55, 57, 90
Sherman, Wyoming, 87
Shurtleff, Roswell, 38, 40
Smith, John, 11, 15–25, 26, 28, 37, 38, 39, 44, 102
Society of Inquiry, 68
Spring, Gardner, 42
S.S. *The Charles A. Young*, 82
Stiles, Ezra, 14
Surveyor's chain, 6, 9

Telescope, 13, 77
Thayer School of Engineering, 110
Ticknor, George, 24, 25
Trustees, Board of, 39, 42, 46, 47, 49, 54, 60, 63, 65, 70, 78, 88, 93, 110
Tucker, William Jewett, 96, 113, 119
Tyler, Bennett, 42

United States Coast and Geodetic Survey, 82
United States Coast-Lake Survey, 69
United States Congress, 63, 64
United States Military Academy, 82
United States Supreme Court, 40

Vacuum tubes, 99
Vermont Domestic Missionary Society, 60

Welch, J. B. G., 101
Wentworth, John, 12, 14
Wentworth, Paul, 12
West, Katherine, 114
Western Reserve, 69
Westford Academy, 85
Wheelock, Eleazar, 2, 4, 5, 11, 12, 17, 18, 19, 37
Wheelock, Eleazar, Jr., 14
Wheelock, James, 14

Wheelock, John, 2, 4, 11, 14, 26, 35, 37, 39, 40
Wheelock, Mary, 2
Wilder, C. T., 109
Wilder Hall, 97, 99, 109, 115, 119, 123
Williams College, 53, 96
Woodward, Bezaleel, 1–14, 25, 26, 34, 35, 37, 44, 101, 102

X Rays, 102, 103, 104, 105, 106

Yale University, 2, 96
Young, Ammi B., 47, 51
Young, Charles A., 54, 60, 68–84, 86, 87, 101, 107, 109, 117, 122
Young, Dyer B., 47
Young, Ira, 45–56, 57, 68, 101, 102